THE GREENPEACE BOOK OF
ANTARCTICA

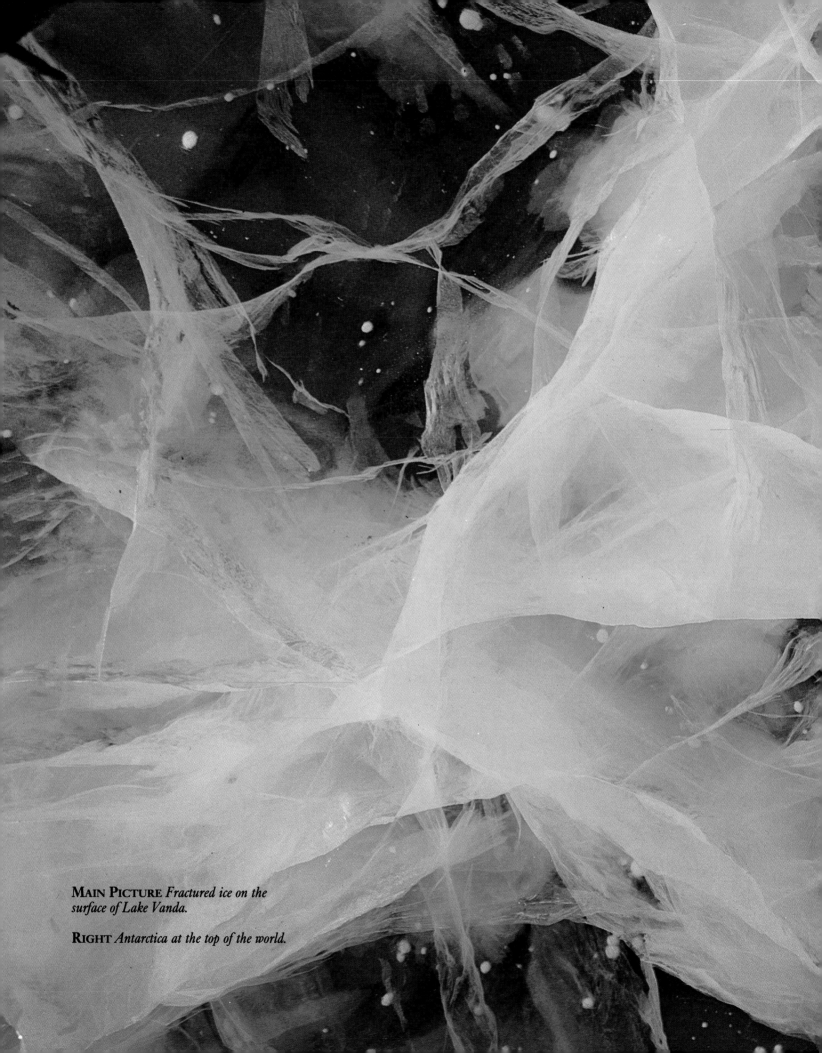

MAIN PICTURE *Fractured ice on the surface of Lake Vanda.*

RIGHT *Antarctica at the top of the world.*

THE GREENPEACE BOOK OF
ANTARCTICA

A new view of the seventh continent

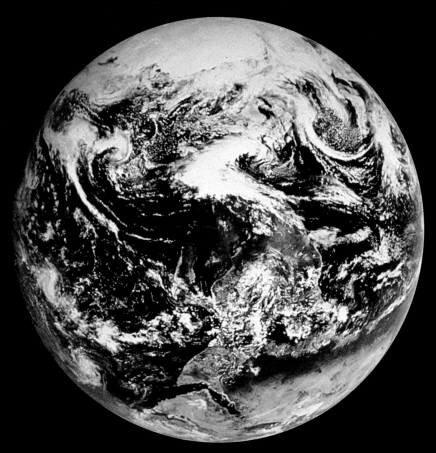

JOHN MAY

DORLING KINDERSLEY
LONDON

GREENPEACE BOOKS
Editorial Director *John May*
Editorial Production *Ian Whitelaw*
Editorial Assistant *Tanya Seton*

DORLING KINDERSLEY
Art Editor *Alex Arthur*
Project Editor *Jane Laing*
Designers
  Sarah Kellaghan
  Sandra Archer

Art Director *Roger Bristow*
Editorial Director *Jackie Douglas*

Consultants
  Bob Headland
  Charles Swithinbank

Text Contributors
  Martin Baker
  Maj de Poorter
  Sir Peter Scott

Main Photographic
Contributors
  Doug Allan
  Colin Monteath
  Eliot Porter

First published in Great Britain in 1988 by
Dorling Kindersley Limited, London

British Library Cataloguing in Publication Data
May, John, 1950–
    The Greenpeace Book of Antarctica.
    1. Ecology—Antarctic Regions
    I. Title II. Greenpeace
    508.98'9   QH84.2

    ISBN 0–86318–283–6

Typeset by MFK Typesetting Limited
Reproduced by Spectrum Reproductions Limited,
Colchester, England
Printed and bound in Spain by Artes Graficas, Toledo S.A.

D. L. TO: 1639 -1988

FOREWORD

SIR PETER SCOTT

I first went to Antarctica in 1966. I sat at the table in my father's over-wintering hut at Cape Evans, where he was photographed in 1911, before setting off on his 800-mile (1,300km) journey to the South Pole. My father and his four companions died on the return from the Pole, when I was two years old, but I have always been aware of the strong family connections with the Antarctic.

Since my first visit, I have been back four times, and each visit has underscored the dangers which that unique and breathtakingly beautiful wilderness faces today. Even before my father's time, commercial sealers and whalers had begun the exploitation of Antarctica's immense stocks of seals and whales, which only stopped as species after species became so depleted that it was uneconomical to hunt them any more. Today it is the fish stocks that are being exploited, and it seems only too possible that krill, the foundation species in the Antarctic ecosystem, will suffer next from man's inability to control his greed.

As we make our mark on more corners of the Earth, it is becoming ever more important to save what remains unspoilt. Even though most people will never have the opportunity of seeing for themselves the amazing ice formations, the vast penguin colonies or the awe-inspiring views of the mountains and glaciers of Antarctica, it is still a great consolation to know that somewhere on Earth there exists a whole continent that is an almost pristine wilderness. Constant vigilance is necessary to make sure that Antarctica remains as untouched as possible, for isolation has been a poor defence against contamination from the world. Adélie penguin eggs have been found with traces of DDT, and plastic rubbish is often washed ashore.

One of the most serious threats hanging over Antarctica is the prospect of the exploitation of its oil and minerals if commercial deposits are found underground or beneath the sea. The Antarctic Treaty Consultative Parties, of whom there are now 20, are discussing how, rather than whether, this should happen. They acknowledge that there are very real risks of oil spills, pollution, and increased human occupation of the few ice-free areas but they seem to regard these risks as perfectly acceptable in exchange for a few years' additional supply of raw materials.

The urge to find out more about remote parts of the Earth drove my father to the Antarctic. That is very different from the urge to overcome technological challenges in order to exploit all the Earth's resources. I believe we should have the wisdom to know when to leave a place alone. *Peter Scott.*

CONTENTS

ABOVE *The aurora australis in April.*
MAIN PICTURE *A view across the upper Wright Valley, one of the dry valleys in Victoria Land.*

LIFE AT THE END OF THE WORLD

ABOVE *A sooty albatross nesting on South Georgia.*
MAIN PICTURE *Delicate lichens encrust a seemingly inhospitable rock at Port Lockroy.*
BELOW *Chinstrap penguins ride on a huge iceberg.*

Ice crystals growing in an ice tunnel.

INTRODUCTION

A vast, cold zone of whiteness, Antarctica is a blank space on which generations of men and women have projected their ambitions, hopes and fears. Indeed, the story of man's exploration of this cruel, yet beautiful continent, of the heroic struggle between man and the extreme environment, is well documented. **The Greenpeace Book of Antarctica** takes a fresh look at this fascinating continent. Written for all those people who, like the author, will never go to Antarctica but who none the less want to visit this extraordinary region in their imaginations, it seeks to develop a new attitude to the seventh continent, one in which preservation of the environment becomes the central consideration.

Partly a popular encyclopedia, partly an account of Greenpeace's campaigns in the region, this highly informative book represents the work of writers, photographers, scientists and campaigners from all over the world. It shows many of the splendours of this remote and remarkable ice wilderness, strengthening the case for its conservation.

This is a time of increasing world interest in Antarctica. As Alan K. Herikson remarked, "Antarctica is suspended between two ages – those of competitive nationalism and cooperative internationalism. It thus is not only a place in geography but also a phase in history. No other part of the world so clearly exhibits the backward-looking and the forward-looking aspects of today's international system."

Greenpeace's campaign to establish the region as a World Park, to minimize the effects of the human presence there and to prevent the exploitation of the area's mineral reserves, is gaining increasing support as more people become aware of Antarctica's fragile nature. Its extreme vulnerability to change, highlighted by modern scientific research, makes it a sensitive monitor of the changes that humans are making to the biosphere.

The Greenpeace base in front of Mount Erebus.

This alone would be reason enough for its complete preservation. Yet Antarctica should also be preserved simply because it represents for us all a place of pristine whiteness. In an increasingly polluted world, this clean and largely unspoilt area should on no account be lost.

This book stands at the point where conservation meets imagination. I hope that it enriches both causes.

John May, LEWES

TERRA INCOGNITA

"... *the desire to wipe out* terra incognita
*appeals more deeply to certain instincts of human
nature than either science or trade.*"
Hugh R. Mill, polar geographer (1909)

"*I felt as though I had been plumped upon
another planet or into another geologic horizon of
which man had no knowledge or memory.*"
Admiral Richard E. Byrd – *Alone* (1938)

ANTARCTICA EVOLVING

The fifth largest continent, covering 10 per cent of the Earth's land surface, Antarctica stretches across 14 million sq km (5.4 million sq mls).

A cold, windswept desert, with a climate drier than the Sahara, more than 99 per cent of Antarctica's area is covered in a layer of ice estimated, on average, to be 2,500m (8,000ft) thick. This means that Antarctica has the greatest average height of any continent.

Climatic conditions can be extreme. A record low temperature of −89.2°C (−126°F) was recorded at the Soviet Vostok base on 21 July 1983; a wind speed of 88m per sec (200 mph) was measured at the French Dumont d'Urville base in July 1972.

In the beginning

Antarctica has not always been a frozen continent. Seventy million years ago, at the dawn of the Cenozoic era, the climate was probably semi-tropical. The land was covered with deep forests and populated by land animals about which we as yet know little. The oceans abounded with giant reptiles like plesiosaurs and with bony fishes. For it is now believed that Antarctica once formed the nucleus of the Gondwana supercontinent, with South America, Africa, Australasia and India grouped all around it.

Continental drift

It was on 6 January 1912 that Alfred Wegener first proposed the notion that the continents were not fixed in stable positions but were constantly moving relative to one another. Another 40 years passed before there was enough scientific evidence to prove him right.

Wegener's theory of continental drift has now been developed into the science of plate tectonics. According to this complex science, the surface of the Earth is composed of a series of "plates", which move about on convection currents in the Earth's molten

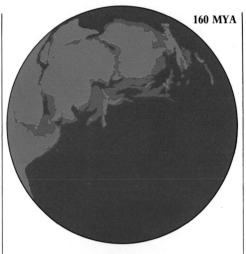

160 MYA

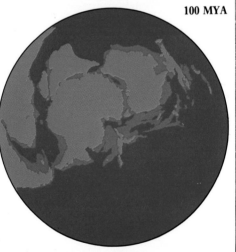

100 MYA

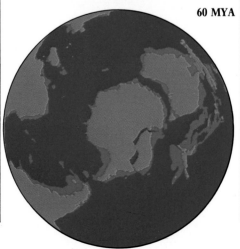

60 MYA

THE CHANGING FACE OF ANTARCTICA

These computer graphic reconstructions are centred on the South Pole at the time indicated. The present land is shown in green and the continental shelves, which correspond roughly to the present 200m (650ft) submarine contour, in light blue.

Antarctica was over the South Pole 280 million years ago when it was part of the Gondwana supercontinent. At about this time the supercontinent began to move north and the climate became much warmer. Volcanic eruptions were common 175 mya, and by 160 mya the supercontinent was beginning to break up. Madagascar (with India, Australasia and Antarctica) started to separate from East Africa. Then, 140 mya, South America started to move away from Africa and the South Atlantic Ocean opened

India, Australasia and Antarctica continued to move relative to Africa 120 mya. By 115 mya Madagascar stopped moving away from East Africa. At 100 mya Australasia and Antarctica began to separate.

At 60 mya Australasia and Antarctica moved more strongly away from one another, and the circumpolar current began to surround Antarctica. A cool-temperate climate prevailed in Antarctica, giving rise to the growth of vegetation and trees. At 25 mya Arabia started to separate from Africa, and the Red Sea began to open. At this point, Antarctica became covered in ice. It now occupies almost exactly the same position over the South Pole as it did 280 mya.

The major sources of scientific uncertainty in the images are: 1) The timing of the break-up of the components of Gondwana; 2) The shapes of the continents at the time, particularly Greater Antarctica and New Zealand; 3) Where the individual parts of Greater Antarctica were prior to the present.

interior. Although their movement is slow (perhaps 1cm/½in a year), over millions of years they cause continents to collide, break up and reform.

Geological jigsaw puzzle
The exact original fit of the continents that once formed the Gondwana supercontinent is still a matter of scientific speculation, with the greatest controversy centring around the exact genesis of Lesser Antarctica.

Geologists are exploring the theory that this area is made up of a number of crustal fragments, each of which has moved separately. It appears that they have not only moved laterally but may also have rotated over time, thus making the reconstruction of geological events a complex jigsaw puzzle.

Scientists have established that the Gondwana supercontinent existed from at least 500 million years ago (Cambrian period) to about 160 mya (late Jurassic period). They believe that it split apart in several stages, starting at about this time when Madagascar (with India, Australasia and Antarctica) started to separate from East Africa.

The last to move away from Antarctica was Australasia. They began to separate about 60 mya ago or so, and Antarctica – now an island continent – became surrounded by the chill water of the circumpolar current. Temperatures dropped at a dramatic rate. At about 25 mya the forests disappeared and Antarctica became covered with ice.

Scientific evidence
The first direct evidence to support the theory that the ancestors of Australia's present-day marsupials came from Antarctica rather than North America, as was previously thought, was found by US scientists in 1982. They discovered fossils of a small, berry-eating marsupial, dating from 40 to 45 mya.

In 1987, scientists from the United States and New Zealand discovered the complete skeleton of a 40 million year-old killer whale. The bones, which weigh more than 1.5 tonnes, reveal a 10m- (30ft-) long creature with a 1.2m- (4ft-) long skull and jaws lined with 10cm- (4in-) long triangular teeth.

On Seymour Island, a barren area situated close to the top of the Antarctic Peninsula, were found the remains of a 1.8m- (6ft-) tall, flightless bird together with the jaw of a crocodile.

In December 1985, ancient marine plants and fossilized wood were discovered at high altitudes in the Transantarctic Mountains. In the same year, Australian scientists discovered a fossilized dolphin on Prydz Bay.

These recent discoveries have been identified as between 2 and 5 million years old, suggesting that the climate then was 20°C (36°F) warmer than today. They also suggest that the ice sheet may have melted and reformed many times and that streams and lakes existed at the land surface situated below where the South Pole is today.

Exciting discoveries are still being made and evidence accumulated to build up a full picture of the past – and continuing – evolution of Antarctica.

FOSSILS
The earliest discoveries in Antarctica were of the seed fern *Glossopteris*, which have also been found in Africa, South America and India.

Another key discovery was that of specimens of the land reptile *Lystrosaurus*, a creature whose remains have also been found in other continents. The fact that they, and other related animals, were incapable of crossing oceans, makes their widespread distribution difficult to explain by any means other than that of continental drift.

Fossils found in Antarctica include stumps of coniferous trees, sharks' teeth, remains of turtles, fish and whales, and four fossil penguins, including one 1.5m (4ft 11in) tall.

An island continent

Antarctica is now separated from the other continents by a fluctuating barrier of floating pack-ice and by the stormy Southern Ocean. South America is 1,000km (600mls) away across the roughest stretch of waters in the world. More distant still are the coast of Australia (2,500km/1,550mls away) and the tip of Africa, which lies 4,000km (2,500mls) distant.

Scattered around the Southern Ocean are a number of small islands and archipelagoes on both sides of the Antarctic Convergence, which is generally taken to be the boundary of the Antarctic region.

THE SOUTHERN HEMISPHERE

This polar stereographic projection has been composed using data from the NOAA satellite. It shows summer conditions in January in the southern hemisphere. Clouds and snow appear white, green vegetation appears red and other terrain is in shades of grey and blue.

The Convergence is a belt of water about 40km (25mls) wide, where cold, northerly flowing currents sink beneath warmer, circulating waters. It marks not only a change in the ocean's surface temperature but also in its chemical composition. This affects the marine and bird life, which is different on either side of this natural boundary. The area north of the Convergence is known as the sub-Antarctic.

Antarctic waters

Currents and often tempestuous winds circulate around the Southern Ocean, south of the Convergence, and transport huge tabular icebergs. There is pack-ice over most of the southern portions in winter, although it varies greatly in amount and position from year to year. However, it is surprisingly productive. During the months with sunlight, food chains develop quickly, and phytoplankton blooms support rapidly increasing numbers of krill – the key organism in the marine cycle.

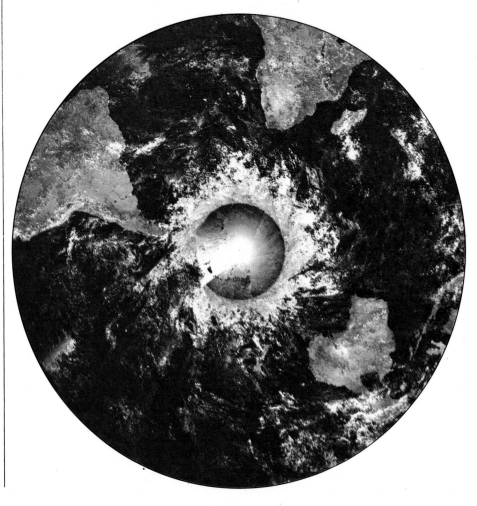

ANTARCTIC ISLANDS

The majority of islands in the Southern Ocean lie close to the Antarctic Convergence, where the mixing of different water masses creates cold, wet and windy weather. Most of these islands are geologically young and volcanic in origin but terrain varies greatly. They provide ideal breeding sites for a wide range of seabirds and mammals.

SOUTH GEORGIA

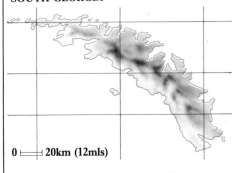

0 |———| 20km (12mls)

This large island of 3,755km² (2,332mls²) was an important base for 19th-century sealers, and was the largest Antarctic whaling centre between 1904 and 1966.

HEARD ISLAND

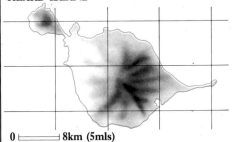

0 |———| 8km (5mls)

Ninety per cent of this island is covered in glacial ice. Seal populations were devastated by sealers who came here from the mid-1800s until about 1910.

SOUTH ORKNEY ISLANDS

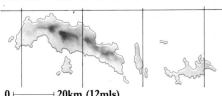

0 |———| 20km (12mls)

This isolated, mountainous archipelago covers an area of 622 km² (386 mls²). A meteorological observatory was established here in 1903.

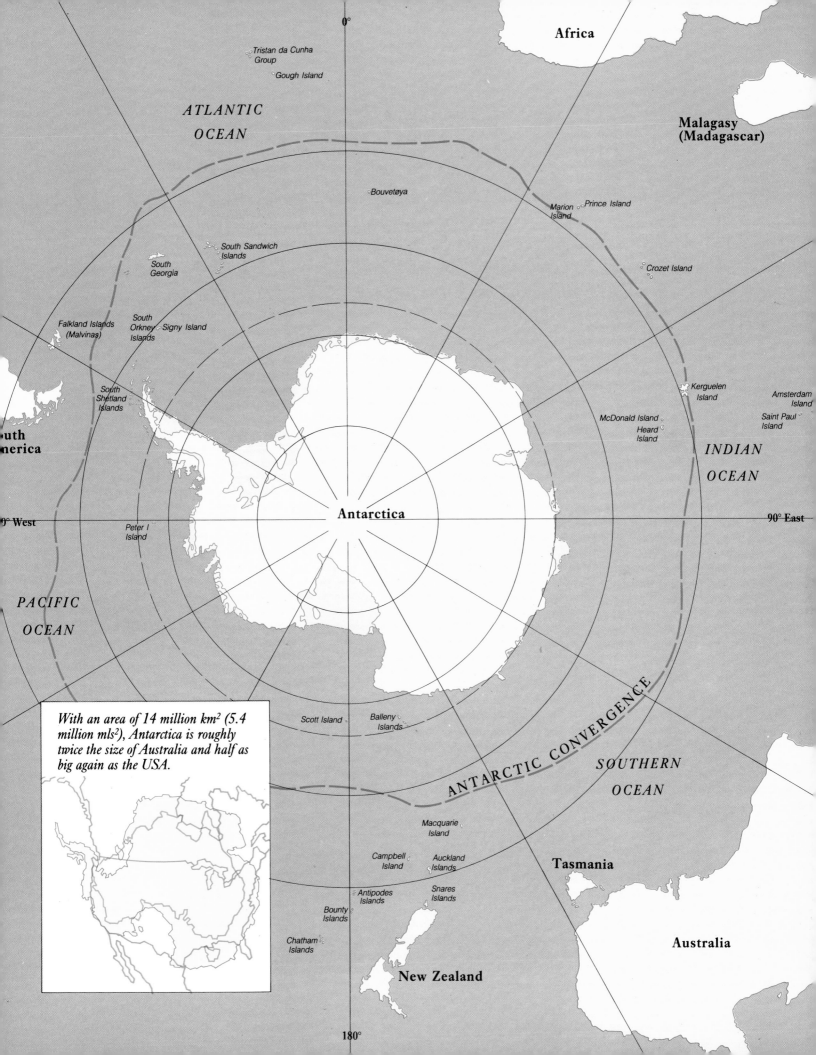

0°

Tristan da Cunha
Group

Gough Island

Africa

Malagasy
(Madagascar)

ATLANTIC

OCEAN

Bouvetøya

Marion
Island

Prince Island

South Sandwich
Islands

Crozet Island

South
Georgia

Falkland Islands
(Malvinas)

South
Orkney
Islands

Signy Island

Kerguelen
Island

Amsterdam
Island

South
Shetland
Islands

McDonald Island

Heard
Island

Saint Paul
Island

INDIAN

OCEAN

uth
nerica

Antarctica

0° West

90° East

Peter I
Island

PACIFIC

OCEAN

Scott Island

Balleny
Islands

ANTARCTIC CONVERGENCE

SOUTHERN

OCEAN

*With an area of 14 million km² (5.4
million mls²), Antarctica is roughly
twice the size of Australia and half as
big again as the USA.*

Macquarie
Island

Campbell
Island

Auckland
Islands

Tasmania

Antipodes
Islands

Snares
Islands

Bounty
Islands

Australia

Chatham
Islands

New Zealand

180°

ICE SHEETS

A vast, interlocking mass of grounded ice sheets and floating ice shelves, Antarctica contains 90 per cent of the world's ice and most of the planet's fresh water reserves. This immense volume of ice is estimated at roughly 30 million cu km (7 million cu mls), and is formed from the compacted accumulation of some 100,000 years of snow. If the ice sheets were to melt, the world ocean would rise as much as 60 to 65m (200 to 210ft).

The continent is divided into two distinct regions – Greater and Lesser Antarctica – and the bulk of the continent's ice, some 26 million km^3 (6.3 million mls^3) is contained in the Greater Antarctic Ice Sheet. This enormous ice dome rises steeply from the coast and flattens to a plateau in the interior, where it reaches a maximum height of just over 4km (2.5mls) above sea-level.

Surprisingly, the thickest ice is located only 400km (248mls) from the coast in Terre Adélie, where a huge, sub-glacial trench is filled with 4,776m (15,670ft) of ice. Some of the thinnest ice is found near the centre of the continent, where a sub-glacial mountain chain rises to a maximum height of 3,500m (11,500ft). Here, isolated peaks of rock, or nunataks, peep through the ice. The Bentley Subglacial Trench marks the lowest point on the continent at 2,538m (8,325ft) below sea-level.

EXPOSED ROCK ON THE COAST
Of the five per cent of Antarctic coastline that is free of ice, the majority is to be found on the Peninsula (above).

MOUNTAINS OF ICE
The cross-section (below) runs from Ronne Entrance in Lesser Antarctica to Colvocoresses Bay in Greater Antarctica.

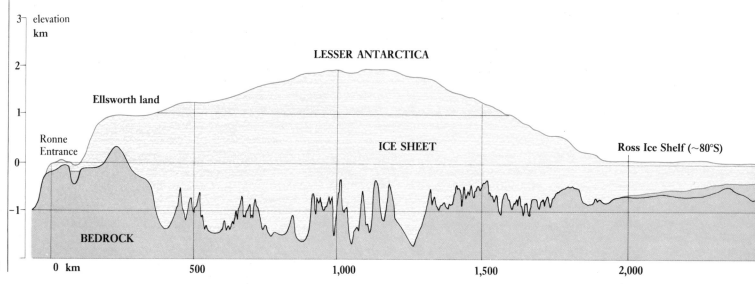

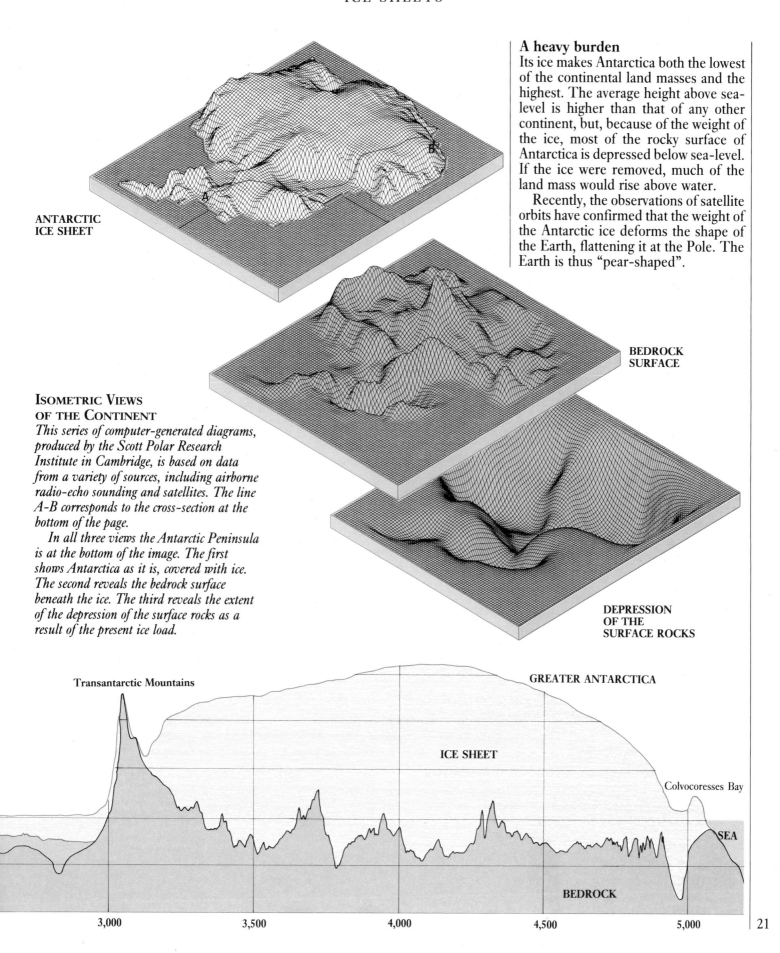

**ANTARCTIC
ICE SHEET**

A heavy burden
Its ice makes Antarctica both the lowest of the continental land masses and the highest. The average height above sea-level is higher than that of any other continent, but, because of the weight of the ice, most of the rocky surface of Antarctica is depressed below sea-level. If the ice were removed, much of the land mass would rise above water.

Recently, the observations of satellite orbits have confirmed that the weight of the Antarctic ice deforms the shape of the Earth, flattening it at the Pole. The Earth is thus "pear-shaped".

**BEDROCK
SURFACE**

**ISOMETRIC VIEWS
OF THE CONTINENT**
This series of computer-generated diagrams, produced by the Scott Polar Research Institute in Cambridge, is based on data from a variety of sources, including airborne radio-echo sounding and satellites. The line A-B corresponds to the cross-section at the bottom of the page.

In all three views the Antarctic Peninsula is at the bottom of the image. The first shows Antarctica as it is, covered with ice. The second reveals the bedrock surface beneath the ice. The third reveals the extent of the depression of the surface rocks as a result of the present ice load.

**DEPRESSION
OF THE
SURFACE ROCKS**

Transantarctic Mountains

GREATER ANTARCTICA

ICE SHEET

Colvocoresses Bay

SEA

BEDROCK

3,000   3,500   4,000   4,500   5,000

SNOW AND ICE

The average annual snowfall over the Antarctic continent is far from evenly distributed. Moisture in the warm air that spirals in from the Southern Ocean cools as it reaches the continent, and falls as snow, mainly around the coast. The snowfall decreases further inland, and the central region receives virtually none. The snow that has formed the ice cap over central Antarctica has taken thousands of years to accumulate. Glaciers flow slowly from the domed centre of the continent to the coast.

SNOWFLAKES
The shapes and sizes of the water crystals that make up snowflakes depend upon the altitude and temperature at which they form (left).

AVERAGE ANNUAL SNOWFALL
This map of the annual accumulation of snow over Antarctica shows that the central region is a virtual desert (below).

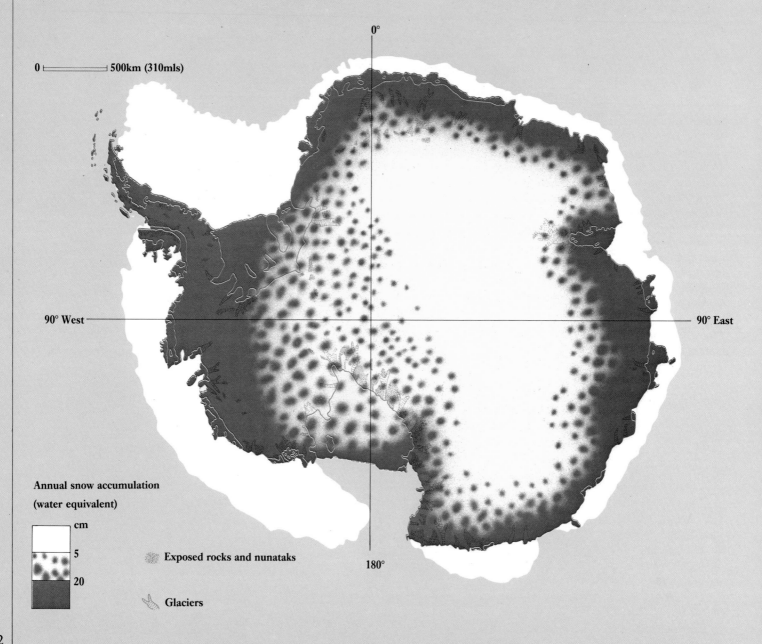

0 ⊢———————⊣ 500km (310mls)

0°

90° West

90° East

180°

Annual snow accumulation

(water equivalent)

cm

5

20

Exposed rocks and nunataks

Glaciers

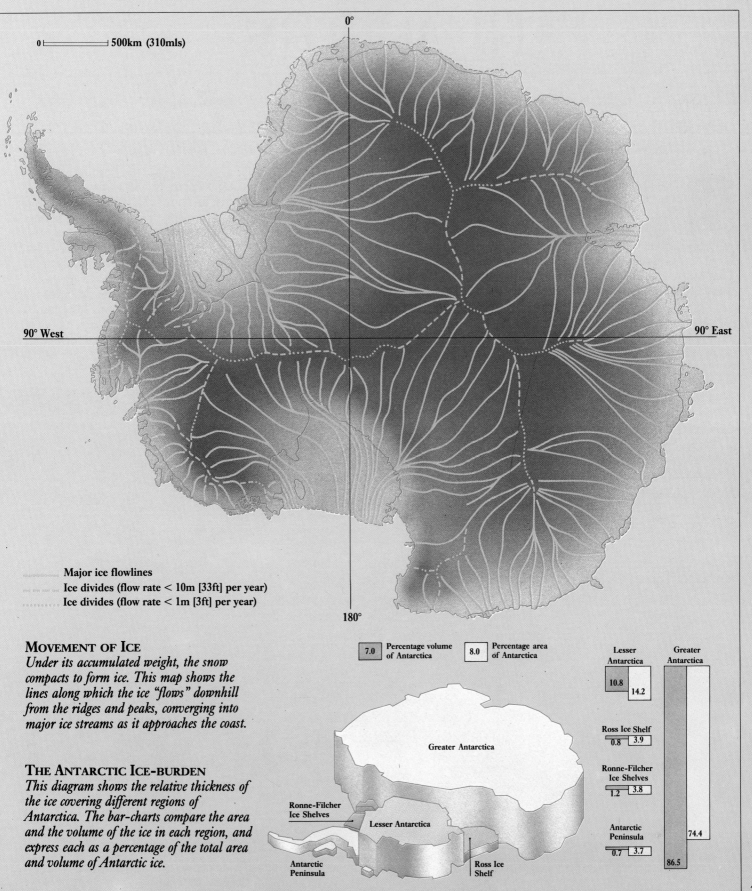

0°

0 ⊢━━━━━━┥ **500km (310mls)**

90° West

90° East

━━━━━━ **Major ice flowlines**
━ ━ ━ ━ **Ice divides (flow rate < 10m [33ft] per year)**
••••••••• **Ice divides (flow rate < 1m [3ft] per year)**

180°

MOVEMENT OF ICE
Under its accumulated weight, the snow compacts to form ice. This map shows the lines along which the ice "flows" downhill from the ridges and peaks, converging into major ice streams as it approaches the coast.

THE ANTARCTIC ICE-BURDEN
This diagram shows the relative thickness of the ice covering different regions of Antarctica. The bar-charts compare the area and the volume of the ice in each region, and express each as a percentage of the total area and volume of Antarctic ice.

7.0 Percentage volume of Antarctica

8.0 Percentage area of Antarctica

Greater Antarctica

Ronne-Filcher Ice Shelves

Lesser Antarctica

Antarctic Peninsula

Ross Ice Shelf

Lesser Antarctica — 10.8 / 14.2

Greater Antarctica — 86.5 / 74.4

Ross Ice Shelf — 0.8 / 3.9

Ronne-Filcher Ice Shelves — 1.2 / 3.8

Antarctic Peninsula — 0.7 / 3.7

GLACIERS

"The glacier surface was like an ocean in torment. As far as the eye could see, serrate ridges and giant furrows were aligned parallel with the abrupt rock walls of the valley. Between them lay endless fields of crevasses." (C. Swithinbank at the Byrd Glacier)

Seven major valley glaciers drain into the Ross Ice Shelf: the Byrd, Mulock, Nimrod, Beardmore, Shackleton, Scott and Amundsen. The Byrd Glacier is the largest: almost as wide as the English Channel, it discharges more ice annually than all the others combined.

Layer upon layer of snow
Essentially a mass of densely packed snow, glaciers take hundreds of years to form. The snowflakes fall on the surface of the ice sheet in the form of hexagonal crystals but, within days, their delicate extremities disappear. Gradually, the snow crystals become rounded, forming into a layer of grains, trapping air pockets between them.

As more snow falls, weighing down heavily on this layer, the air spaces become smaller and smaller until the grains cannot be packed down any more. Under continuing pressure, due to the weight of the snow above, the grains of snow change their size and shape, allowing the air spaces to become even smaller and more isolated and making the snow less permeable to air. This consolidated snow is known as névé or firn. Gradually all the air spaces are closed off and the material becomes impermeable. At this point it becomes true glacier ice. Ninety-five per cent of the ice sheet of Antarctica is formed of such glacier ice.

THE LAMBERT GLACIER
The world's largest valley glacier, the Lambert Glacier, drains about a quarter of the ice mass from the Greater Antarctic ice sheet. It flows for over 400km (250mls) through the Prince Charles Mountains, discharging more than 35km³ (8.4mls³) of ice into the Amery Ice Shelf every year, where it is nearly 200km (125mls) wide.

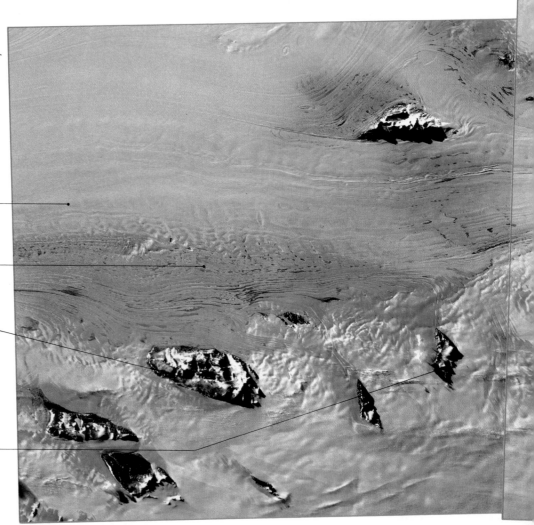

AMERY ICE SHELF

ICE RUMPLES

FISHER MASSIF

SHAW MASSIF

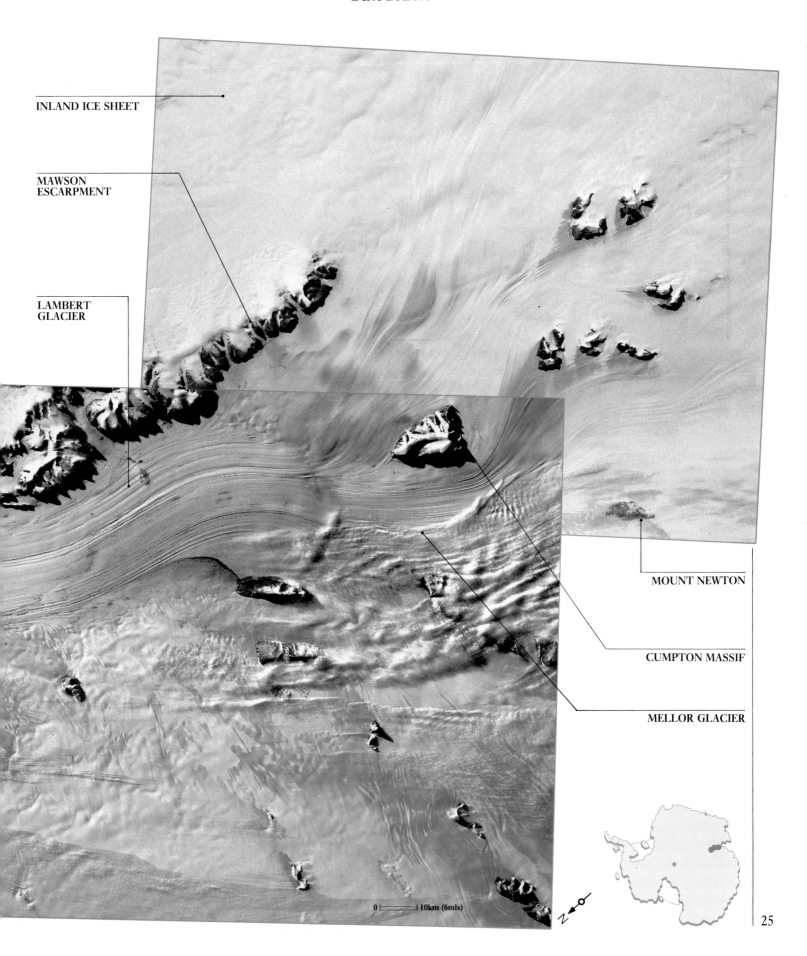

INLAND ICE SHEET

MAWSON
ESCARPMENT

LAMBERT
GLACIER

MOUNT NEWTON

CUMPTON MASSIF

MELLOR GLACIER

0 ⊢———┤ 10km (6mls)

THE BYRD GLACIER

This Landsat image, taken on 16 January 1974, shows the Byrd Glacier at its junction with the Ross Ice Shelf. The Byrd is the largest glacier flowing from the Transantarctic Mountains, and it discharges 18km³ (4.3km³) of ice into the Ross Ice Shelf every year.

Glaciologist, Charles Swithinbank, stood on the left bank of the glacier, overlooking the ice. He comments: "We could follow the flow lines sweeping uninterrupted through 80 miles from the high plateau on our right to the ice shelf on our left. The scale of the whole landscape was appalling."

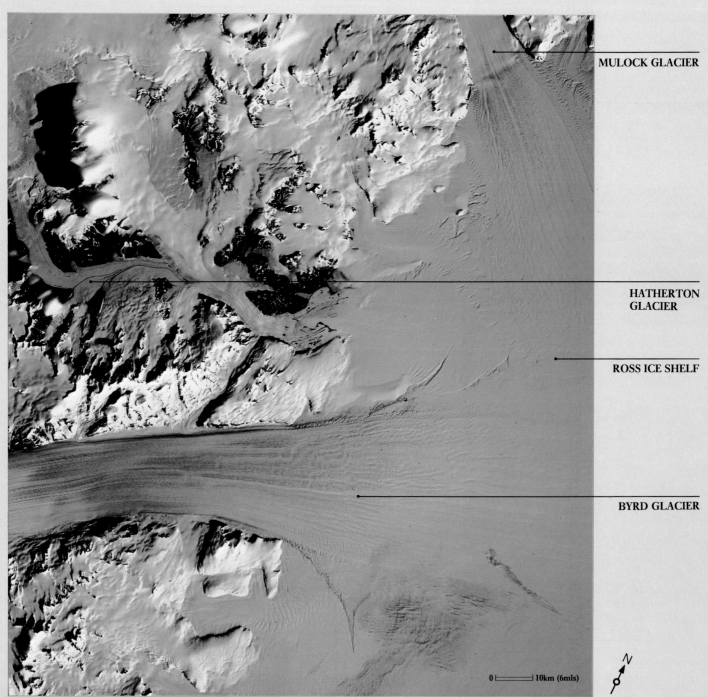

MULOCK GLACIER

HATHERTON GLACIER

ROSS ICE SHELF

BYRD GLACIER

0 ⊢———⊣ 10km (6mls)

N

ANALYSIS OF A GLACIER

In 1968, scientists from the US Army Cold Regions Research and Engineering Laboratory drilled a core hole in a polar glacier down to its bedrock at Byrd Station. Photographed in polarized light, these pictures show cross-sections of that core at different depths.

1 *At a depth of 56m (180ft), the firn has just turned into ice. Air bubbles trapped among the snow grains comprise approximately 10 per cent of its volume. The concentration is in the order of 100 to 150 bubbles per cm³ (1,600 to 2,500 bubbles per in³).*

2 *At 1,105m (3,600ft), the size of the grains has increased and the size of the air bubbles has decreased because of compression under the ice load. At this depth, the air bubbles have virtually disappeared because the pressure has forced the grains to absorb gas molecules.*

3 *At 2,183m (7,160ft), which is 23m (75ft) above the ice bed, the grains have become very large, owing to the increased temperature and pressure.*

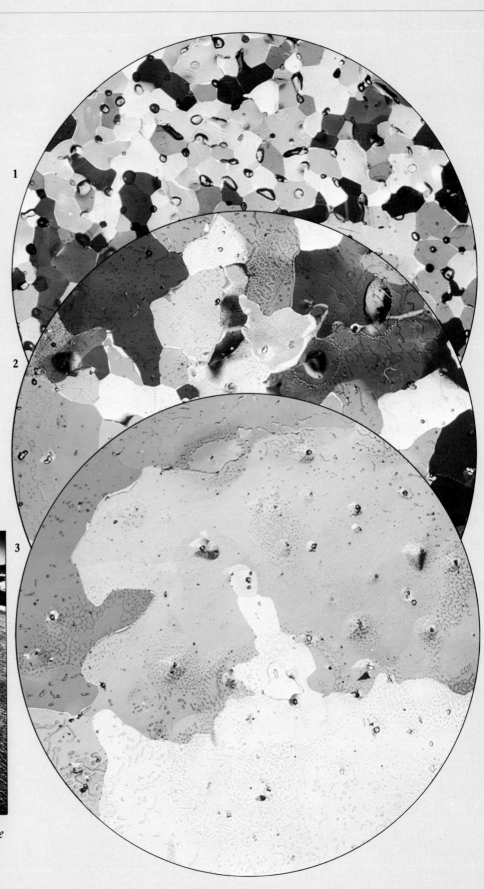

LOOKING UPSTREAM *This aerial image gives some indication of the enormous scale of the Byrd Glacier.*

Firn is less white than snow and it becomes translucent blue or green as it turns into glacier ice. The transition takes place at increasing depths the further inland you travel. At the South Pole, glacier ice forms at a depth of about 100m (300ft) in snow about 1,000 years old; on the Ross Ice Shelf, the transition occurs at 35 to 60m (120 to 200ft) in snow 200 to 300 years old.

Rivers of ice?

Athough these huge masses of accumulated ice flow outwards under their own weight they do not flow like water. In fact, the term "rivers of ice" is a complete misnomer. The rate of flow depends on the size of the ice body, its thickness, the slope of its surface and the temperature of the ice, especially on its bed. Two types of movement are at work: basal sliding and internal deformation, or creep. Most glaciers combine the two as they advance.

Basal sliding is the movement of the glacier over its own bed due to gravity. This can only occur if there is a thin film of water between the ice and the bedrock to reduce friction. If it is too cold for any water to exist, the glacier remains frozen to its bed, and can only move by internal deformation. This movement follows the laws of physics that govern the bending of iron or rock. So glaciers would be better described as "rivers of rock".

Under the tremendous weight and pressure – 300 tonnes per m^2 (30 tonnes per ft^2) at depths of 3,000m (10,000ft) – the ice crystals rearrange themselves in layers of molecules more or less parallel to the surface of the glacier. These layers then begin gliding over one another. Glaciers are thus dynamic features, fresh snow falling on the top of the glacier, the pressure pushing the ice down and forwards, and material being lost, particularly at the snout, through ablation (melting and sublimation). In the Antarctic, most glaciers are stable because, although the snowfall is sparse, little is lost through ablation. If this equilibrium is upset, the snout of the glacier will advance or retreat.

Cracks and crevasses

The flow varies within an individual glacier. The beginning and the end generally move faster than the centre, and there are different flow rates between various levels of the ice. This variation creates stresses in the glacier, resulting in vast cracks or crevasses.

The cracks range in size from a few millimetres to over 30m (100ft) across, and occur mainly in the brittle top crust.

At depths of 30m (100ft) the ice is under so much pressure that it deforms plastically and does not crack. Crevasses form across the direction of the ice flow due to sudden changes in the slope of the underlying bedrock, which result in the glacier speeding up. Longitudinal ones occur wherever a glacier is advancing at more than one speed. Most glaciers contain both sorts of crack, to make a criss-cross pattern. Traversing glaciers is therefore a somewhat dangerous manoeuvre, especially as the cracks are often concealed by – often fragile – snow bridges.

Monitoring the glaciers

Scientific understanding of glaciers was revolutionized by the development of radio-echo sounding from aircraft. This method is much faster and more effective than ground-based soundings of the ice, using seismic charges.

Not only could the thickness of the ice be measured, but also the internal layering of the ice owing to changes in density and electrical conductivity. Sub-glacial mountain ranges and even substantial lakes were discovered hidden beneath the ice sheet.

Satellite monitoring has also enabled glaciologists to measure the speed of ice movement more conveniently.

CREVASSES
The variation in flow rates between different parts of the glacier creates stresses in the ice which produce crevasses (left). They make travelling across the ice an extremely dangerous operation as their presence is often concealed by bridges of snow that are sometimes thick enough to hold a person, and sometimes not!

SASTRUGI
The strong winter winds of Antarctica combine with the heavy coastal snowfall to produce a dramatic landscape. Together they erode the surface, sculpting it into a corrugated "sea" of irregularly shaped snow forms, known as sastrugi (right).

ICE SHELVES

Only about five per cent of Antarctica's coastline is composed of rock. The vast mass of the interior ice sheet meets the shore in the form of massive, awe-inspiring ice shelves and impregnable grounded ice cliffs.

Ice shelves are floating slabs of land ice that rise and fall with the ocean tide and form "a kind of crystalline exoskeleton that shapes, binds, and protects the ice continent" (Stephen J. Pyne).

The critical boundaries of an ice shelf are the grounding line, or hinge, where the glacier or ice sheet begins to float, and the ice front, where the shelf disintegrates into icebergs. There is a constant balance between ice flow from the continent, ice added by snowfall, and bottom freezing; and ice is lost by bottom melting and the calving of icebergs. The stability of the ice shelves rests on maintaining this equilibrium.

Anchoring the ice shelves

Where the ice shelf passes over rocky shoals below sea-level, it forms dome-shaped ice "rises" or low ice "rumples".

These act like spot welds on a car, anchoring the ice shelf in place. Without these pinning points, the inland ice sheet could push some of the ice shelves out to sea.

These anchors are also a source of instability, however, as they create fractures and stresses within the shelves. The surface is gouged by the wind into the ridge and trough icescape called "sastrugi", and riven by deep crevasses; its interior ice is infiltrated in places by brine. All this is covered by a blanket of snow to give the appearance of a calm, flat, untrammelled white land.

The Great Ice Barrier

The Ross Ice Shelf is the largest of all the ice shelves. Roughly the size of France, or twice the size of New Zealand, it alone accounts for 30 per cent of all Antarctic shelf ice. It was first sighted in 1841 by Sir James Clark Ross, who called it the Great Ice Barrier. For two successive seasons he sailed along its edge, looking for an opening to the south, or a landing place, but finding neither. Up to 900m (3,000ft) thick at its grounding line, and sloping to 200m (650ft) at the ice front, it is constantly replenished by five ice streams and numerous glaciers.

Other major ice shelves include: the Amery, which has formed where the Lambert Glacier empties into an enormous fjord; the Filchner Shelf, largely sustained by the Slessor and Recovery glaciers; and the Ronne Shelf, supported by the Foundation and Evans ice streams.

ICE CLIFFS   *An Antarctic scientist is dwarfed by the enormous cliffs of ice off the Brunt Ice Shelf (above). The jagged ice cliffs north of Rothera (right) are no less awe-inspiring.*

GEORGE VI ICE SHELF

This Landsat image, taken on 9 January 1973, shows the George VI Ice Shelf, which lies between Alexander Island and the English Coast of Palmer Land. The image is 180km (110mls) across, and the area of open water is the Ronne Entrance to the Bellinghausen Sea.

Below George VI Sound, warm deep water spills up over the continental shelf, producing water temperatures almost 2°C (3.6°F) higher than those found under the Ross Ice Shelf. This may explain the fact that the melt rate of the George VI Ice Shelf is 10 times greater than that of the Ross Ice Shelf.

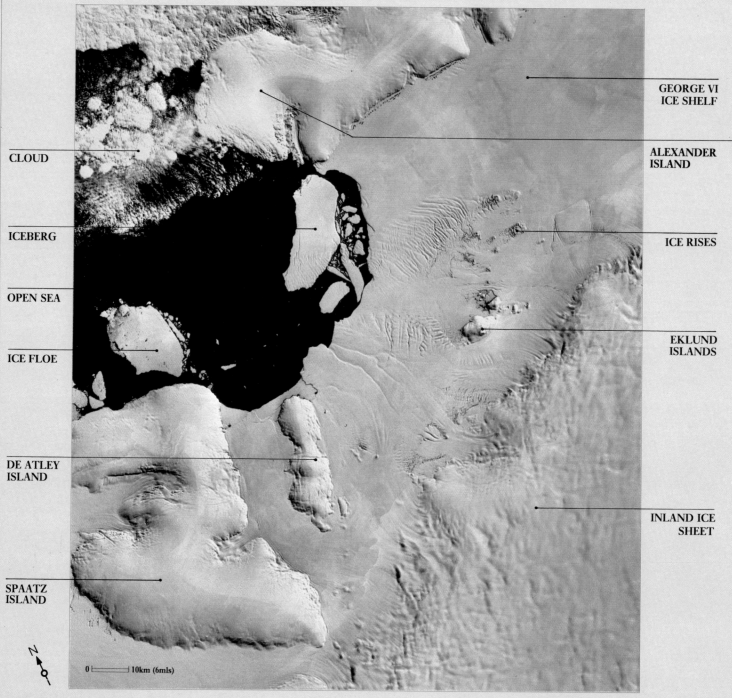

GEORGE VI ICE SHELF

ALEXANDER ISLAND

CLOUD

ICEBERG

ICE RISES

OPEN SEA

ICE FLOE

EKLUND ISLANDS

DE ATLEY ISLAND

INLAND ICE SHEET

SPAATZ ISLAND

0 ⊢——⊣ 10km (6mls)

Ekström Ice Shelf

This satellite image highlights the jagged edge of the ice shelf, which is 30m (100ft) above sea-level. It also shows the vehicle tracks and cargo recently unloaded from the ice-breaking research vessel, *Polarstern*, close to the West German Georg von Neumayer station on the Ekström Ice Shelf, Dronning Maud Land.

First mapped by the Norwegian–British–Swedish Antarctic Expedition (1949–52), the ice shelf was named after Bertil Ekström, who drowned when his vehicle plunged over the ice edge and through the pack-ice in 1951.

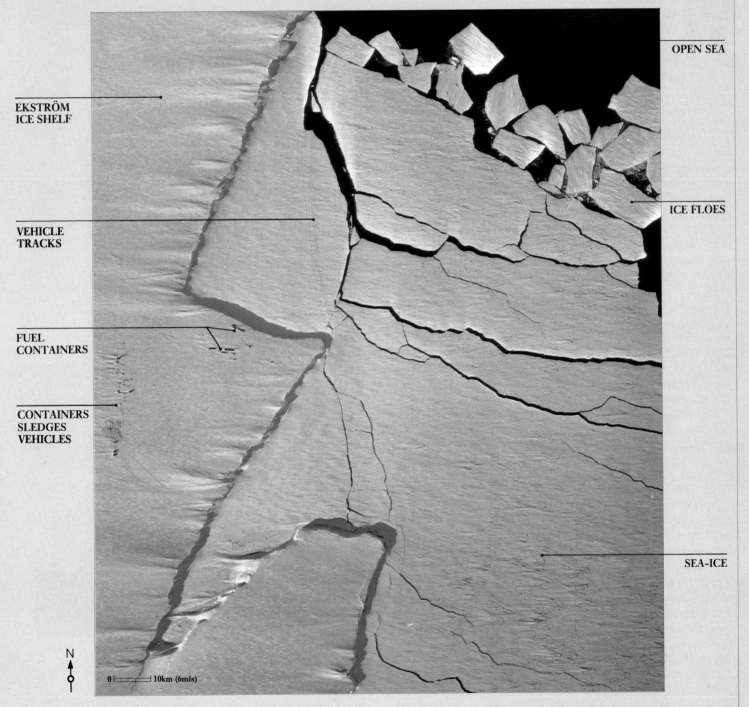

EKSTRÖM ICE SHELF

OPEN SEA

VEHICLE TRACKS

ICE FLOES

FUEL CONTAINERS

CONTAINERS SLEDGES VEHICLES

SEA-ICE

N

0 ⊢——— 10km (6mls)

ICEBERGS

Infinite in variety, widely distributed, ever-changing in shape, size and colour, icebergs have long inspired and terrified the voyagers of the Southern Ocean.

Various estimates are given as to their numbers. A survey in 1965 counted 30,000 in an area of 4,400km² (1,700mls²) between longitude 44°E and 168°E. It is also claimed that more than 1,450km³ (348mls³) of icebergs are calved from Antarctica each year – equivalent to about half the world's water usage.

About four-fifths of icebergs are calved from the ice shelves that make up 30 per cent of the total coastline of continental Antarctica. These are large, sometimes enormous, tabular bergs, usually measuring 200 to 300m (650 to 1,000ft) thick. They are flat-topped with sharply angled sides, gradually weathering into more rounded bergs. One of the largest icebergs ever recorded was sighted by the *USS Glacier* on 12 November 1956, 240km (150mls) west of Scott Island. It covered an area of 31,000km³ (150mls³) and was 335km (208mls) long and 97km (60mls) wide – the size of Belgium. Smaller, more irregular, icebergs are calved from the grounded ice cliffs and glaciers.

Icebergs often take years to melt but are riven by internal fractures as they age, breaking into smaller ice forms.

The movement of icebergs

Once free of the parent ice, the bergs drift north and are carried east to west by the circumpolar currents at a speed of around 13km (8mls) a day. Many are later trapped in bays or inlets.

Those that remain free eventually reach the Antarctic Convergence, which reverses their course and sweeps them from west to east. This is the outer limit for most icebergs, now a fraction of their original size, but occasionally a storm or violent eddy pushes them even further north. The most northerly iceberg remnant recorded was sighted by the *Dochra* on 30 April 1894 in the Atlantic at 26° 30′S, 25° 40′W.

CALVING FROM THE ICE SHELF
Huge chunks of ice (above) are calved from the continent wherever the ice sheet meets the ocean. This dramatic photo shows the process in action.

SWARMS OF ICEBERGS
As a large berg drifts away from the ice sheet into warmer waters, it splits and breaks up into hundreds of smaller bergs (left).

CAVES OF ICE
As icebergs disintegrate, they take on some fascinating shapes. The erosive forces of ocean and winds combine to carve holes in the ice along lines of weakness.

TYPES OF ICEBERG

Icebergs can be classified into three main categories: tabular, irregular and rounded. Tabular bergs are young icebergs, not long separated from the ice sheet. They are flat-topped and horizontal, still revealing the original stratification of the ice. As they age, they may become uneven, domed, tilted or blocky.

Irregular bergs are more eroded, with predominately angular features, although some, like the drydock, also have troughs. Many have pinnacles and some look very like pyramids. Jagged and castellate bergs with battlement-like profiles number among this group.

Few angular surfaces remain on the oldest, rounded bergs. Some of these bergs become unstable and flip over to reveal the convex underside. By this stage, much of the original iceberg has melted and they ride low in the water.

TABULAR BERG
This enormous, flat-topped, horizontal tabular berg was sighted in the Weddell Sea. It is five times longer than it is tall.

ICEBERG FORMS *A classification of iceberg shapes (below).*

TABULAR BERGS	original stratification normally visible on close inspection; no evidence of rollover	
horizontal		top surface flat and horizontal; stratification visible and parallel to waterline
uneven		uneven top surface; top may be sloping or undulating; crevasses may be present
domed		top surface slopes down on all sides
tilted		area of original tabular surface greater or equal to half the waterline area
blocky		maximum height greater or equal to one-third width or one-fifth length

ROUNDED BERGS	water-smoothed, convex surfaces predominate	
sub-rounded		some angular concave surfaces (may have complex though smooth topography)
rounded		highest angular feature less than half maximum height
well-rounded		concave surfaces very minor; highest cliff at waterline less than one-third height

IRREGULAR BERGS	angular or irregular features predominate; water-smoothed shapes are dominated by concave surfaces	
tabular remnant		original tabular surface and stratification are visible
pinnacled		one or more pinnacles are present
pyramidal		like a pyramid
drydock		low area in middle often awash
castellate		battlement-like skyline
jagged		sawtooth or jagged skyline
slab		table-like
blocky		maximum height greater or equal to one-third width or one-fifth length
roof		shaped like a roof or A-frame
rounded		generally irregular but rounded surfaces conspicuous
combinations		tabular remnant, tilted with rounded (wave-shaped) platform

MONSTROUS BERGS
Three US Navy ice-breakers attempt to move a 244m- (800ft-) long iceberg (above) – without success. Minutes later, the wind changed direction and the vast berg drifted away unassisted. Although rather small, the craggy pinnacled berg (left) still dwarfs the scientists in the foreground.

BREAKOUT

Shortly after these Landsat images were taken, in early 1986, the Filchner Ice Shelf became 13,000km² (5,000mls²) smaller when all the ice north of the giant chasm calved into the Weddell Sea, taking the closed Argentine base Belgrano I and the Soviet summer base, Druzhnaya, with it. Belgrano I was abandoned in January 1980 after succumbing to the pressures of ice, but Druzhnaya (meaning "friendly") was due to be reoccupied in December 1986. Instead, the Soviet expedition spent several weeks early in 1987 searching the Weddell Sea for their missing base, which they found buried under deep snowdrifts on an iceberg in February that year. The ice had split into three giant icebergs by then.

Unexpectedly, a simultaneous breakout of ice occurred from the Larsen Ice Shelf on the opposite side of the Weddell Sea. The two combined represent the biggest calving of ice in recorded history, amounting to approximately 6,000km³ (1,440 mls³) of ice. This enormous amount of ice is equivalent to three times the average annual ice loss from the whole of the Antarctic ice sheet.

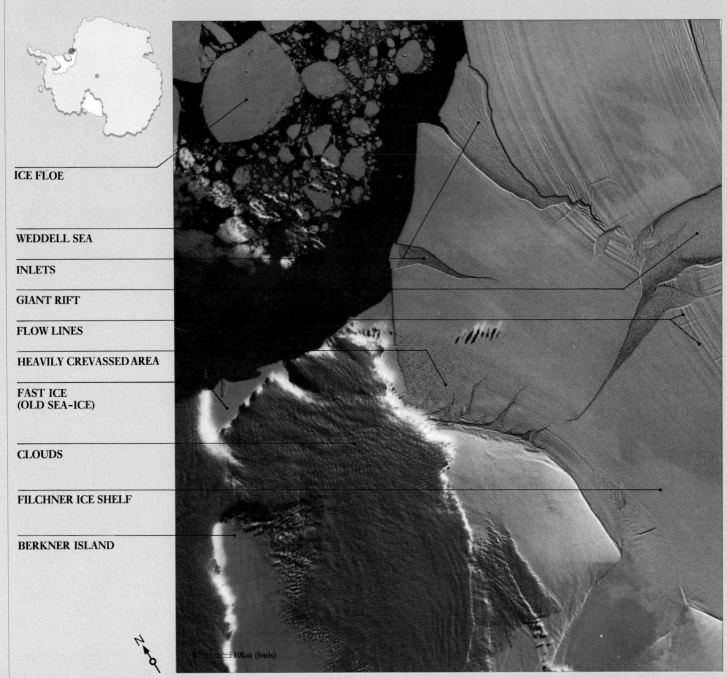

ICE FLOE

WEDDELL SEA

INLETS

GIANT RIFT

FLOW LINES

HEAVILY CREVASSED AREA

FAST ICE
(OLD SEA-ICE)

CLOUDS

FILCHNER ICE SHELF

BERKNER ISLAND

0 ⊢⎯⎯⎯⎯⎯⊣ 10km (6mls)

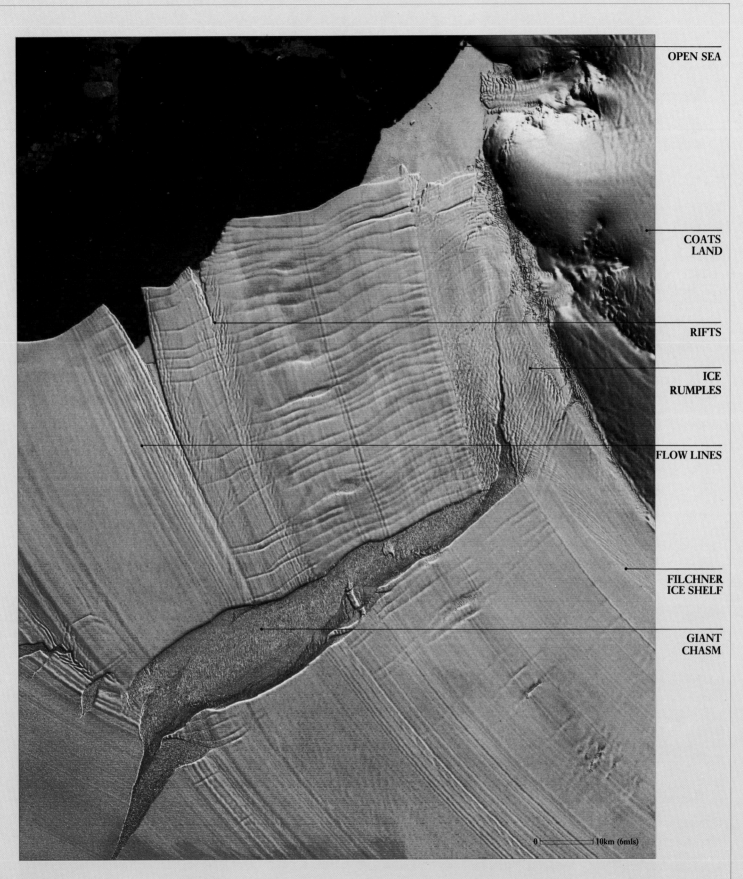

OPEN SEA

COATS
LAND

RIFTS

ICE
RUMPLES

FLOW LINES

FILCHNER
ICE SHELF

GIANT
CHASM

0 |————————| 10km (6mls)

SEA-ICE

Sea-ice, which covers seven per cent of the world's oceans, has a considerable influence on the global atmospheric and oceanic circulation. It reduces significantly the amount of solar radiation absorbed at the Earth's surface, and it restricts the amount of heat transferred from ocean to atmosphere.

Each year the area of the sea-ice around Antarctica increases and decreases in an ancient cycle. It is at its maximum in September (20 million km² / 7,700,000mls²) and at its minimum by the end of summer in February (4 million km² / 1,500,000mls²). The seasonal difference is therefore greater than the whole of the area of the Antarctic continent itself.

On average, the advancing edge of the sea-ice moves 4.2km (2.6mls) a day and the total ice cover increases by about 100,000km² (40,000mls²) a day.

Grease ice and frazil ice

Sea water of average salt content freezes at about −1.8°C (28.8°F), forming first as small, hexagonal crystals on the surface of the supercooled brine. In stable

GREASE ICE *At the end of summer, the ocean begins to cool and the surface starts to freeze, forming grease ice (right).*

CONGELATION ICE
Stained with algae and punctuated with ice stalactites, this underwater view of congelation ice (above) was taken in the sheltered waters of McMurdo Sound.

waters, the crystals become stacked in long needles of congelation ice, or they grow side-by-side in plates, coating the surface with an oily sheen, which is known as grease ice.

Below the surface, particularly in open or turbulent waters, where the ice crystals are constantly jostled, an unstructured slush known as frazil ice forms. Most areas of the sea-ice begin as a composite mixture of grease ice and frazil ice to form a soupy layer.

Pancake ice and pack-ice

As freezing continues, grease ice forms a thin crust, which is usually broken up by wind and waves into separate masses, to form platelets of pancake ice. These platelets, like water lily pads, become curled at the edges as they collide with each other. They then coalesce, acquiring additional layers of frazil ice from below and a covering of snow on top. This thickening layer continues to be broken up by the waves into pack-ice.

ICE SCULPTURES

"Frost flowers" form on sea-ice around nuclei of salt excreted from freezing ice columns (below). In early winter the waves break the grease ice into platelets, which collide with one another to form pancake ice (bottom).

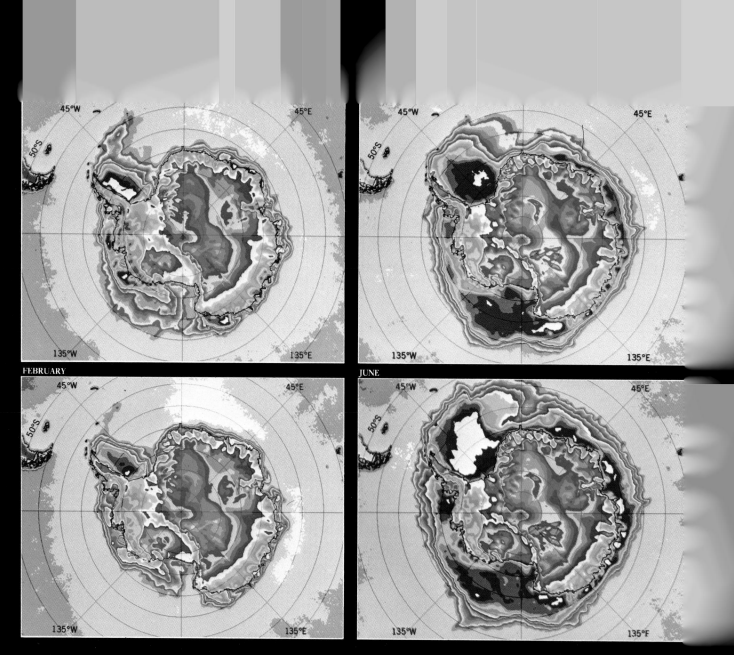

FEBRUARY

JUNE

SEA-ICE CYCLE

This set of satellite images, showing the average sea-ice cover around Antarctica over the four-year-period 1973-1976 is based on data from the Electrically Scanning Microwave Radiometer (ESMR) on the Nimbus 5 spacecraft, which was launched in 1970. The data provided an all-weather, day and night view of the sea-ice cycle for the first time.

The ESMR data were transmitted to two spaceflight tracking stations near Fairbanks, Alaska, and Rosman, North Carolina. They, in turn, relayed it to the Goddard Space Flight Centre, where it was transferred to magnetic tape and processed by computers. The

values were corrected to allow for differing ice temperatures, then plotted on a polar stereographic map, and colour-coded. Three-day average maps were generated and these were used to create monthly, yearly, and multi-yearly, average maps.

Measuring the salinity of sea-ice

The ESMR is a passive device that records the intensity of naturally emitted radiance from the Earth, known as the brightness temperature. The ratio of the brightness temperature to the physical temperature is called the emissivity. The sharp contrast between the microwave emissions from open water and that from sea-ice enables the

brightness temperatures recorded by the ESMR to be converted into sea-ice concentrations.

The most important physical property that distinguishes new ice from first-year ice and multi-year ice is the salinity in the ice above sea-level (the freeboard). During the freezing process, pockets of concentrated brine are trapped between fine layers of ice plates. The amount of brine trapped depends on the temperature during formation and the age of the ice. The brine content decreases over time either as a result of heating, as a result of the surface melt-water percolating through the ice, or by drainage due to gravity. Each kind of ice has a slightly

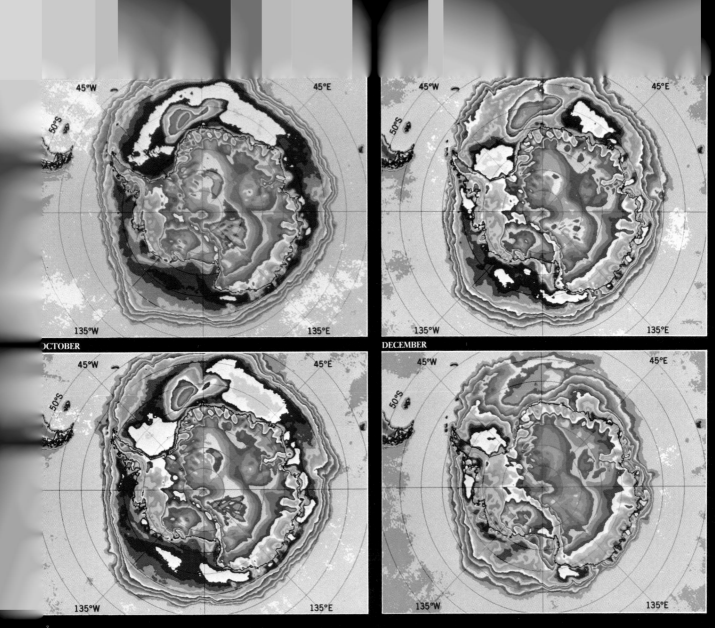

OCTOBER

DECEMBER

different radiation signature, so a certain amount of discrimination can be made between ice types.

Analysing the data

Several interesting discoveries were made as a result of analysing the ESMR data. First, there was a steady shrinking of the average total area of sea-ice during the four-year period. This is now known to be an anomaly. Sea-ice cover increased from 1966 to 1972 and again from the mid-1970s.

Second, it was observed that the rate of melting of the ice pack during the spring was faster than the growth rate in the autumn, an asymmetrical cycle not found in the Arctic. This may be caused by the upwelling of relatively warm deep water, adding heat to the atmosphere-to-ocean heat flux.

Third, large and shifting areas of open water (polynyas) were discovered within the ice pack (coloured pink in these images). They cover a much greater area than was thought.

Weekly sea-ice charts

Before the artificial satellite era, our knowledge of sea-ice movements was limited to isolated observations from ships and aircraft. Now weekly sea-ice charts are compiled using data from a number of satellites. In 1978, the SEASAT satellite carried a synthetic aperture radar (SAR), an active system which transmits its own signal and receives the reflection from the ground. It gives greater image detail and a number of new satellites carrying it are planned. They may be able to provide ships with details of sea-ice conditions almost instantly.

All the evidence so far obtained from satellites shows that the Antarctic sea-ice is subject to a wide range of complex cycles, from region to region, from season to season and from year to year. These are linked to changes in atmospheric and oceanic circulation patterns. However, these data cover a relatively short period of time and scientists are cautious about extrapolating present trends into the future.

BENEATH THE ICE

The geographical division of the Antarctic continent into Greater and Lesser Antarctica is also the geological one with the Transantarctic Mountains separating the two.

Beneath the ice, Greater, or East, Antarctica is an old, stable platform or "shield" of Precambrian rocks (more than 570 million years old). It is overlain on its western margins with layers of marine sediments from 240 to 590 mya.

These sediments constitute much of the Transantarctic Mountains and, because the area has been geologically stable, they are arranged in horizontal

ANCIENT CRYSTALLINE ROCKS
The Framnes Mountains in Kemp Land near Mawson Coast are part of the ancient continental shield of Greater Antarctica.

layers, a phenomenon that is known as layer-cake geology.

A great mountain chain
The Transantarctics are one of the world's great mountain chains, stretching from Victoria Land on the Ross Sea to Coats Land on the Weddell Sea, a total distance of 4,800km (3,000mls). They have developed along a major rift that separates the ice sheets of Greater and Lesser Antarctica. In many places, the range is virtually submerged beneath ice, with only isolated nunataks, or peaks, poking through. In addition, it has been extensively shaped by glaciers.

THE GEOLOGICAL DIVISION
An ice-free view of Antarctica (above). The Olympus Range (right) forms part of the Transantarctic Mountains.

ELLSWORTH MOUNTAINS

South of the Peninsula, in Marie Byrd Land, lie the Ellsworth Mountains, named after Lincoln Ellsworth, the millionaire US aviator who discovered their northern portion, the Sentinel Range, during his transcontinental flight in 1935.

Although the Ellsworth Mountains are in Lesser Antarctica, they consist almost entirely of sedimentary rocks of the same age as those that make up the Transantarctic Mountains; however, their structuring is at right angles to the rocks of the Transantarctics and they are more deformed. This has led geologists to the conclusion that they are a fragment of the Earth's crust that has been rotated.

The highest peaks in this range are also the highest peaks in Antarctica – the Vinson Massif (4,897m/16,000ft), first climbed on 17 December 1966, and Mount Tyree (4,723m/15,500ft).

THE SENTINEL RANGE *Looking northwards along the Ellsworth Mountains, Mount Tyree lies at the most western point.*

Lesser Antarctica

Lesser, or West, Antarctica is half the size of the eastern portion of the continent and includes the Peninsula and Marie Byrd Land.

It has a turbulent geological history. Originally it formed part of the Pacific margin of Gondwana, which began to break up 170 million years ago in Jurassic times. Geologists now consider the area to be made up of three major crustal fragments: the Marie Byrd Land block, separated from the Thurston Island block by the Byrd Basin (more than 2,000m/6,500ft deep in places); and the Peninsula itself, separated from the Thurston Island block by another deep trough. These blocks are welded together and attached to the main continental mass by the overlying ice cap. Radar-sounding techniques are now being used to try to discover whether these deep depressions are major geological structures or merely sub-glacial valleys scooped out by ice streams.

THE ANTARCTIC PENINSULA *The view from Lully, Alexander Island (right), and the Lemaire Channel in the Antarctic Peninsula (below).*

The Antarctic Peninsula itself consists of Mesozoic (65 to 225 million years ago) and Tertiary (2 to 65 mya) rocks and younger glacial deposits. The mountain-building processes responsible for the Andes were the same as those that formed the mountains of the Peninsula. The two ranges are connected by a submarine ridge, which includes active volcanoes in the South Sandwich Islands. The South Orkneys and South Georgia are also peaks of this chain. Their geological structure has been continually "reset" by intrusions of igneous rocks formed by the interaction of two crustal plates.

THE OASES

By far the largest continuous areas of ice-free land in Antarctica are the oases, a questionable term for desert-like areas of low precipitation and freezing temperatures, where only highly specialized algae and lichens can survive.

Collectively, the oases account for only a small part of the total surface of Antarctica, but they are scientifically important. Small coastal oases are found at Cape Hallet in north Victoria Land, near the site of the Australian Mawson station in MacRobertson Land, at the Bunger Hills in Wilkes Land and in the Vestfold Hills in Princess Elizabeth Land.

The Dry Valleys

The most extensive oases are the Dry Valleys in Victoria Land. Lying in the shelter of the Royal Society Range covering an area some 15 to 25km by 150km (9 to 15mls by 93mls), they consist of three main valleys that were once occupied by glaciers. They are the Taylor Valley, the Wright Valley and the Victoria Valley.

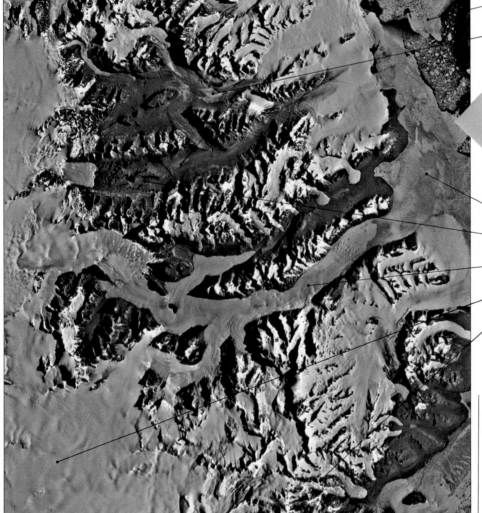

PACK ICE

LAKES

FAST ICE

DRY VALLEYS

FERRAR GLACIER

INLAND ICE SHEET

ROYAL SOCIETY RANGE

THE MCMURDO SOUND REGION
In this false-colour, multi-spectral digital mosaic of Landsat imagery (above) Ross Island is the snow-covered area to the right centre of the image, topped by the active volcano Mount Erebus (left centre of island). The Dry Valleys on the south coast of Victoria Land can be more clearly seen in the enlargement (left).

DRYGALSKI ICE TONGUE

ICE FLOE

PACK ICE

McMURDO SOUND

MOUNT EREBUS

CAPE EVANS

McMURDO STATION

ROSS ICE SHELF

N

0 ⊨▭▭▭⊨ 25km (16mls)

ACTIVE VOLCANO

Mount Erebus (3,794m/12,444ft high) is the largest of four major volcanoes that form Ross Island. When discovered and named by Sir James Clark Ross in 1841 it was emitting smoke and flame to a height of some 600m (2,000ft) above the mouth of the crater. The most significant recorded eruption since that time occurred during the southern, or austral, summer of 1984/85 when gas trapped under the solidified crust of the internal lava lake was released in a series of violent explosions. Some scientists think that the volcanoes erupting towards the head of the Ross Ice Shelf 20 million years ago were instrumental in the uplift of the Transantarctic Mountains.

DESERT LABYRINTH

The Wright Valley (left) is typical of the inhospitable Dry Valleys. The mummified seal carcass (right) may have been preserved here for some 3,000 years.

No rain has fallen here in the last two million years at least. Once free of the glacial ice, these oases became self-perpetuating. The exposed surface absorbs enough heat from the sun to evaporate the winter snows and the cold, dry, gravity-driven winds (katabatic) blow off any snow that remains.

The Dry Valleys are considered the nearest equivalent on Earth to the landscape of Mars. As a result, NASA carried out a great deal of preparatory work here before launching its Viking probe to the rainless deserts of Mars.

Abstract sculpture and saline lakes

They are particularly interesting places in which to study the forces of erosion, as the valley floors are littered with curiously shaped rocks known as ventifacts. These are like pieces of abstract sculpture: on one side, the surface is hard and shiny, polished by sand-blasting winds; on the other, it is soft and crumbly, weathered into pits.

The other feature of the Dry Valleys is the saline lakes – Lake Vanda in the Wright Valley and Lake Bonney in the Taylor Valley – which are fed for a few weeks each year by streams originating from the snouts of melting glaciers.

LAKE VANDA – GIANT SOLAR HEATER

Lake Vanda is 75m (250ft) at its deepest point and, like Lake Bonney, is for most of the time covered with a layer of ice some 4m (13ft) thick. Despite this, the temperature *increases* the further down you go, reaching a maximum of 25°C (77°F) at the lake bottom. This amazing phenomenon occurs because the ice crystals in the ice layer act as light pipes, transmitting sunlight to the water below. They do this because they are aligned vertically as a result of the layer of ice constantly freezing underneath and evaporating from the top. Lake Vanda acts like a giant natural solar heater.

These beautiful crystals form the structure of the ice in Lake Vanda.

The lake surface water down to 50m (170ft) is drinkable but it becomes increasingly saline at greater depths. This is due to evaporation. Each time water evaporates from the top of the lake, all the dissolved salts are forced to the bottom where they form a dense brine. Fresh water does flow in at the top, but the difference in density is too great to allow the waters to mix to any real degree, except very slowly, through diffusion.

The fresh water inflow comes from the Onyx River, the only one of any size in Antarctica. The Onyx arises from a coastal glacier and flows inland for over 30km (20mls) each summer, replenishing the lake.

WEATHER

Antarctica is the coldest and driest continent on Earth, ringed by the stormiest ocean. The elevated central plateau is a cold desert with annual mean temperatures of between −50°C and −60°C (−58°F to −76°F). Annual precipitation is the equivalent of between only 3 and 7cm (1.25 and 3in) of water. The lower altitude coastal regions are much warmer (−10°C to −20°C / +14°F to −4°F) and receive more rain and have stronger winds.

The continent is extremely cold despite the fact that the total annual number of daylight hours received in the region is about equal to that received in equatorial latitudes. The total radiation received is much less because the sunlight strikes the Poles at a more oblique angle. Furthermore, most of the incoming short-wave radiation from the sun is either absorbed by gases in the atmosphere or reflected into space by clouds and by the snow and ice cover.

Layers of air

The atmosphere above the continent is stratified into three layers, which are only tenuously connected. At the surface and in the stratosphere, cold air is transported away from the continent. In between is a layer of warm air which flows to the Pole from temperate regions. At the Pole it loses its latent heat as the water vapour in it condenses and freezes. If it were not for this input of warm air, the ice cap would become progressively colder, as the ice surface loses more radiation than it receives from the sun.

The windiest continent

Antarctica is the windiest as well as the coldest continent. The land mass is scoured by a regime of persistent and powerful katabatic, or gravity, winds which are the result of cold, dense air rolling down the continental slope from the high plateau.

These gravity winds interact with the warmer air from the ocean to produce a narrow storm belt that rings the continent, producing clouds, fog and extremely severe blizzards.

HOME OF THE BLIZZARD

The sea-ice is being whipped into waves (above) by the extraordinary katabatic winds of Commonwealth Bay on George V Coast. This is the windiest place on Earth, where it is by no means unusual for strong gales of more than 22m per sec (50mph) gusting to 40m per sec (90mph) to persist for days, or even weeks, on end.

CYCLONIC STORMS

This photomosaic of weather satellite images (left) portrays the swirl of cyclonic storms around Antarctica. They circle the continent, gradually spiralling to the coast.

METEOROLOGICAL RESEARCH

Scientific understanding of the key role the Antarctic region plays in the world's weather and climate has been transformed in the last 40 years owing to the use of manned and automatic weather stations, buoys and satellites. Together they provide a greater quantity and quality of information, which enables scientists to build up a much more accurate and complex picture of Antarctica's crucial role in global weather patterns.

Twenty-five meteorological stations operate all-year-round on the continent, most of them in a ring round the coast with two on the high, interior plateau. The gaps in their cover are filled by automatic stations powered by radio-isotope generators, storage batteries or solar cells.

During the Global Atmospheric Research Programnme (GARP) in 1978-79, hundreds of ocean buoys were deployed and the data derived from them correlated with that derived from constant-level balloons.

Even so, the Southern Ocean remains one of the most poorly observed regions on the Earth.

CLIMATE STUDIES *A meteograph is about to be launched in a hydrogen-filled balloon (above). A meteorologist adjusts an automatic weather station (right).*

Further out, beyond the pack-ice, circumpolar westerly winds blow unimpeded round the mid-latitudes of the Southern Ocean, known to sailors as the Roaring Forties, the Furious Fifties and the Screaming Sixties. The large temperature gradients between the cold continent and the warmer ocean continually create areas of low pressure (cyclones). These swirl in a consistent band between latitude 60° and 65°, the atmospheric equivalent of the Oceanic Convergence. They gradually spiral towards the Antarctic coast, where they dissipate. In this way, heat and moisture from the tropics is distributed to the polar region and the energy balance of the globe is maintained.

ANTARCTIC SKYSCAPES
The sun tinges thin nacreous clouds (above) and highlights lenticular clouds (left), producing unearthly effects.

MEASURING PAST CLIMATES BY ICE

To understand fully the current behaviour of our climate and atmosphere we need to know about the climatic past. The Antarctic ice sheet contains a wealth of information in this area because climatic changes are preserved in its layers. To study the layers, scientists drill into the ice to obtain long tubes or ice cores. Some of these contain information about the climate that is as much as half a million years old. The layers of these ice cores are then analysed using physical and chemical techniques.

The "oxygen isotope ratio" is extremely important for studying past climates by ice. Each water molecule contains one atom of oxygen, which weighs between 16 and 18 times as much as a hydrogen atom, depending on the temperature of the air at the time it fell as a snowflake.

Past temperatures

The ratio between oxygen-16 and oxygen-18 rises and falls from winter to summer, and the snow can therefore be dated by counting the seasonal

DRILLING OUT ICE CORES

A scientist removes a 10cm- (4in-) thick ice sample from the core drill. Mechanical drills, using sharp cutters or a heated tip, can remove cores from the ice-sheet in 1m- (3ft-) sections. Two drill holes have been sunk to depths of more than 2km (1.2mls), one of them reaching the bedrock. One such hole was drilled over a period of more than 10 years. When holes are drilled over such a long time they have a tendency to curve, as the ice sheet is always on the move.

Once a core has been removed from the ice sheet (right), analysis can reveal much about the Earth's atmosphere at the time that the ice was formed. It is therefore important to date the ice accurately. This can be done in three ways: in relatively young, shallow ice, the layers of annual deposition can be counted directly; in deeper ice, the decay of naturally occurring elements, such as carbon and chlorine, can be used to indicate the age; and in deep ice, the most commonly used dating method is to work out the flow characteristics of the ice sheet at the core site and then calculate the time required for such characteristics to produce the patterns of layering in the core.

levels (like tree-ring dating). Adjustments to the ratio must be made to account for the fact that the surface elevation of the ice may have changed,. thus causing a change in temperature without a change in the climate.

By analysing the ice cores in this way, the climatic transition from the last ice age to the present interglacial period can be seen clearly. One core, obtained at the Soviet station Vostok in Greater Antarctica, penetrated 2,083m (6,800ft) through ice from the last glacial epoch, the last interglacial period and into the top of ice deposited in the glacial epoch before that.

A dustier world

But it is not only past temperatures that can be ascertained by studying ice cores. By examining deposits in the ice,

scientists have discovered that 18,000-year-old Antarctic ice contains far more dust than present-day snow, and that this dust came from the soils and deserts of other continents. This suggests that there were more arid areas on Earth at that time, together with greater wind speeds capable of carrying the dust such distances.

Danish scientists have developed a technique to measure the electrical conductivity of the ice, which reveals the amounts of volcanic acids that are contained within it.

The greenhouse effect

Another clue to the composition of ancient atmospheres is discovered by measuring the amount of gases like carbon dioxide (CO_2) contained in air bubbles trapped in the ice. Evidence suggests that changes in CO_2 levels could have contributed to the start and finish of ice ages, events perhaps triggered by changes in the amount of incoming solar radiation.

CO_2 concentration has increased by 25 per cent in the last 20 years, owing to the burning of fossil fuels and the destruction of forests. The current interest in the "greenhouse effect" is centred around the rise in CO_2 levels, together with the doubling of methane concentration during the same period. This could lead to a warming of the atmosphere and the melting of the ice sheets with catastrophic results. World temperatures appear to have increased by 0.8°C (1.44°F) during this time and current projections predict a further rise of 2°C (3.6°F) by the year 2030 – enough to alter both agricultural and weather patterns significantly.

The ice cores also show evidence of radioactive material from atmospheric nuclear tests of the immediate post-war period and a four- to 10-fold increase in lead levels. The most distinctive radioactive traces occurred during the austral summers of 1954 to 1955 and 1964 to 1965. These provide an excellent means of determining the mean accumulation rate of snow over a 10-year period.

OPTICAL PHENOMENA

The cold, dry climate and dust-free air of Antarctica mean that there is a complete lack of haze. In such conditions, as many explorers have discovered to their cost, distant objects seem close and mirages are common.

When the air over the polar ice cap is filled with ice crystals, either composed as clouds or simply suspended in the air, the light of the sun and the moon is reflected, refracted, diffracted and scattered by the crystals. This produces what one writer has described as "a

SUMMER SUNSETS
Snow-laden clouds can create dramatic sunsets in Antarctica (below and right).

compendium of light geometry", a varied range of optical effects that have dazzled visitors to the continent.

The exact effect depends on the nature of the crystals in the atmosphere, their orientation to the source of light and the observer's viewpoint. Some, such as halos, or mock suns or moons, and sun pillars, are most dramatically seen in the sky. Others, such as mirages, produce distortions at ground level, causing distant objects to appear close.

HALOS, SUN PILLARS AND FOGBOWS

In the Antarctic skies halos form around the sun or moon in the shape of circles, arcs or spots, known as mock or dog suns or moons. In daylight, sun pillars – vertical streaks of light – can occur. Fogbows are formed by the diffraction of light through super-cooled water droplets, while parhelia are created by sunlight passing through ice crystals. These are called paraselenae if the light is the moon.

Halos, sun pillars and fogbows can occur simultaneously. The explorer, Edward Wilson, observed on a sledging journey "no less than nine mock suns ... and arcs of 14 or more different circles, some of brilliant white light against a deep blue sky, others of brilliant rainbow".

ICE BOW

PARHELION

SUN PILLAR

THE OZONE "HOLE"

A poisonous, highly reactive compound of three oxygen atoms (O_3) – in contrast to the two-atom molecules (O_2) that make up the bulk of the oxygen in our atmosphere – ozone is concentrated into a 50km- (30mls-) thick layer that begins 16km (10mls) above the ground at the lower edge of the stratosphere. It shields life on Earth from the damaging effects of ultra-violet radiation.

The existence of a "hole", or depletion, in this essential protecting layer over Antarctica each spring was noticed by a scientist from the British Antarctic Survey in 1981, and later confirmed by satellite observations. Since then, the size of the "hole" has steadily increased and, in 1987, the layer was half as thick as at the same time of year in 1970.

It is widely accepted that the natural rate at which ozone is destroyed is being increased mainly by the presence of free chlorine atoms which are released by the decomposition of chlorofluorocarbons (CFCs) above the ozone layer. This family of chemicals is used as a coolant in refrigerators and air conditioners, as a blowing agent in the manufacture of foam, and also as a propellant in aerosols. At present, 770,000 tonnes of CFCs are produced every year.

Encouraged by the cold

The damaging effect of chlorine on the ozone layer appears to be enhanced in the Antarctic region by the extreme cold during the three months' darkness of the Antarctic winter. In these conditions, chlorine and other gases accumulate in inert forms.

The ultraviolet light of the spring sun initiates a sequence of complex reactions in which the chlorine and other ozone-consuming gases suddenly become active, a reaction speeded up by the presence of clouds of ice particles. The rapid creation of a "hole" in the ozone layer is the result and this is deepening every year.

In 1987, the American $16 million Airborne Antarctic Ozone Experiment, which included European scientists, made the most detailed study yet of the atmosphere over Antarctica by flying aircraft through the ozone "hole" for the first time.

Alarming implications

This international effort reflects the growing concern among many scientists, environmentalists and politicians that what we are seeing in Antarctica may be the beginning of a much wider breakdown in the ozone layer.

Fears are already being voiced that increased ultraviolet radiation may not only harm humans, but may also have deleterious effects on Antarctic phytoplankton, with far-reaching ecological consequences in the Southern Ocean.

The crucial effect of CFCs on the ozone layer has now been widely accepted, and the first international treaty to limit their production was initially signed by some 40 countries in September 1987. Given that a large proportion of CFCs now present in the atmosphere will still be there in 100 years' time, much more drastic cuts in production will be needed if the ozone layer is to continue to shield the Earth.

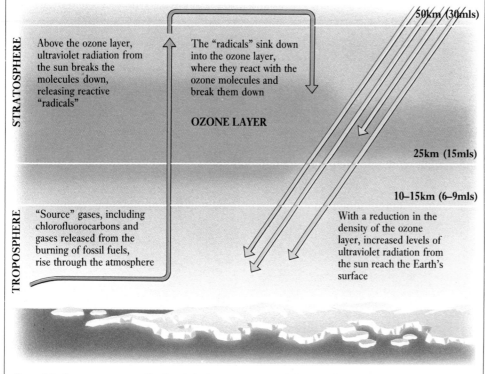

STRATOSPHERE

Above the ozone layer, ultraviolet radiation from the sun breaks the molecules down, releasing reactive "radicals"

The "radicals" sink down into the ozone layer, where they react with the ozone molecules and break them down

OZONE LAYER

50km (30mls)

25km (15mls)

10–15km (6–9mls)

TROPOSPHERE

"Source" gases, including chlorofluorocarbons and gases released from the burning of fossil fuels, rise through the atmosphere

With a reduction in the density of the ozone layer, increased levels of ultraviolet radiation from the sun reach the Earth's surface

Free chlorine atoms, or radicals, sink down into the ozone layer and attack the ozone (O_3) to form chlorine monoxide (ClO) and O_2. The ClO then combines with a free oxygen atom to form O_2 and a chlorine atom. This catalytic reaction continually repeats itself, so that for every chlorine atom released, an estimated 100,000 molecules of ozone are removed from the atmosphere. Other radicals react in similar ways.

FORMATION OF THE OZONE "HOLE"

The total ozone in the southern hemisphere is shown in this map, which illustrates the Antarctic ozone "hole". The map was produced at NASA's Goddard Space Flight Center, using data from the Total Ozone Mapping Spectrometer (TOMS) on board the polar-orbiting satellite Nimbus, which monitors ozone levels over the entire Earth. The ozone "hole" is portrayed in grey and violet colours. It was first observed on 22 August 1986, coloured dark and light grey on this map, and had deepened and expanded by 2 September. By 2 October, the "hole", now coloured violet, had grown to cover an area about the size of the USA, weakening and eventually disappearing two weeks after the final photograph.

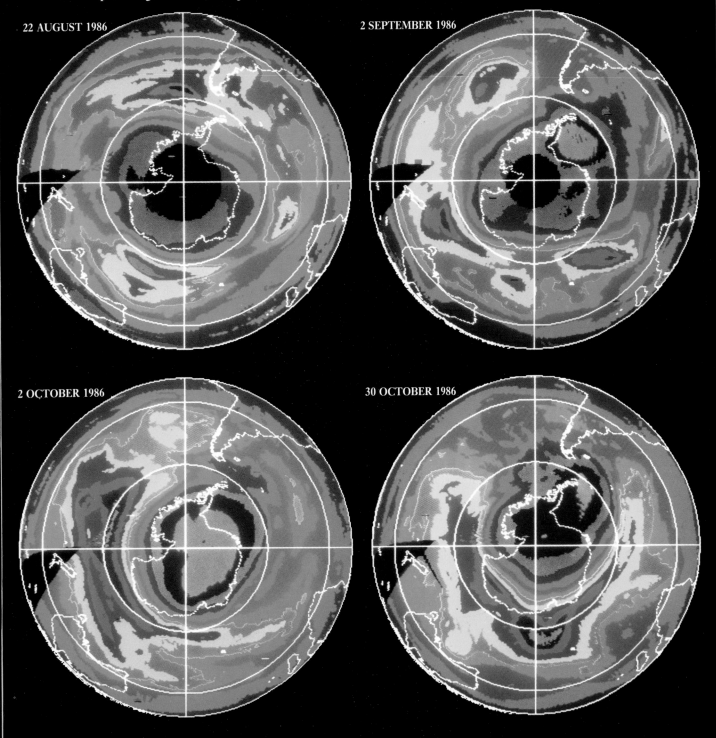

22 AUGUST 1986

2 SEPTEMBER 1986

2 OCTOBER 1986

30 OCTOBER 1986

METEORITES

In the last 15 years, more than 7,500 meteorite fragments have been found on the Antarctic ice sheet. The vast majority of these are stony meteorites, already common in museum collections, but scientists have also found more than 30 fragments of metallic, nickel-iron meteorites, one of which contains minute diamonds – only the second specimen of its kind yet found – formed by impact shock waves.

Some of the meteorites contain traces of at least 20 amino acids, evidence of organic materials that existed in the solar system more than 100 million years before the appearance of life on Earth. In order not to contaminate these valuable fragments, they are handled with stainless-steel shears and tongs, packed in Teflon bags and shipped frozen to laboratories where they are examined in processing cabinets originally designed to test lunar rocks.

Rare findings
The first Antarctic meteorite was discovered by the Australian explorer Douglas Mawson in 1912 but it was another 50 years before Soviet and US scientists discovered three others completely by chance.

Then, in December 1969, a team of Japanese glacial geologists surveying blue icefields inland from the Dronning Fabiolafjella stumbled upon nine black meteorite fragments, which they took to be pieces from a single meteorite shower. Four years later, after laboratory analysis, it was announced that four of the specimens belonged to completely different classes of meteorite. The chances of four different meteorites falling in the same spot are extremely small: they must have been carried to the spot by ice motion.

Trapped in the ice
This would suggest that when meteorites fall on Antarctica they are trapped in the ice – frozen in for thousands or possibly millions of years. The ice cap moves steadily seawards, so most specimens probably reach the coast and

0 |———————| 20km (12.5mls)

MOUNTAINS OF METEORITES
This satellite image shows the Yamato Mountains regions where Japanese scientists have found thousands of meteorite fragments embedded in the ice.

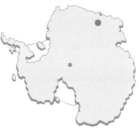

float away in icebergs. However, in some places, where a natural barrier halts the ice flow, winds strip the snow away and the bare ice is eroded. In this way meteorites gradually come to the surface. They are concentrated in small areas by a combination of wind flow and the upward push of the ice.

Japanese scientists returned to the Dronning Fabiolafjella over the next four seasons and joint Japanese-American expeditions began searching icy areas near the Dry Valleys which looked promising on satellite and aerial photos. By 1982 more than 5,000 fragments of several hundred meteorites had been found, an increase of more than 25 per cent on the world total.

Fragments of the moon

In January 1982, far to the west of the main meteorite concentrations, in the Allan Hills region, a 32g (1.1oz) meteorite, the size of an apricot, was discovered in the last few days of a US expedition, and is believed to be a piece of the moon.

Microscopic examination of a small slice of this light-green "lunar" meteorite showed a striking similarity to rocks brought back by the Apollo astronauts from the lunar highlands. It is speckled with white fragments of anorthosite – calcium- and aluminium-rich rocks that make up the bulk of the lunar highland crust and give it the light colour that is visible from Earth.

In order to have escaped from the moon's gravitational field, this piece of rock must have been accelerated to a velocity of more than 2.4km per sec (1.5 mls per sec). The only force believed to be powerful enough to do this is the high-energy impact of a large asteroid.

Rocks from Mars

Two other unusual stony meteorite fragments found in Antarctica are of the same type as seven fragments found in Australia, Brazil, Egypt, France, India, Nigeria and the US; they are all believed to be part of the planet Mars and pose even greater explanatory problems. The escape velocity required would be 5km per sec (3mls per sec) and no material could survive the shock of the impact required without vaporizing.

The best mechanism so far suggested is by researchers at the California Institute of Technology. They argue that the rocks were blasted into space by a meteorite that hit Mars at an oblique angle. The impact would vaporize a large amount of rock but, instead of dissipating in all directions, this hot gas would form a powerful jet with a speed of up to 20km per sec (12.25mls per second), which would have sufficient force to pick up rocks and speed them out of the planet's gravitational pull without evaporating them.

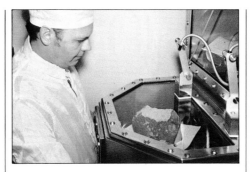

FOUND IN THE ICE
Dr Donald D. Bogard, a geochemist at NASA's Johnson Space Center, studies an Antarctic meteorite, believed to be of Martian origin. The 20kg (45lb) iron meteorite shown below was found in the Allan Hills, Southern Victoria Land.

LUNAR METEORITE
Discovered in the Allan Hills, this 32g (1.1oz) meteorite is believed to be lunar in origin. It has been examined at NASA's Johnson Space Center.

Other research has strengthened the case for the meteorites being of Martian origin. These meteorites are igneous rocks, formed from molten rock or magma, which are substantially younger than all other meteorites – 1,300 million years old compared to the 4,500 million years of most meteorites. Radioactive dating also confirms that they were subjected to shock waves about 180 million years ago. Thus their parent body must have been an object where volcanism was still occurring 1,300 million years ago and where a powerful impact took place only 180 million years ago. Mars is the most likely candidate.

Analysis of the gases trapped in pockets of fused brown glass inside one of the SNC meteorites (named after three of the localities where they fell – Shergotty [India], Nakhla [Egypt] and Chassigny [France]) has revealed traces of argon, neon, krypton and xenon in similar proportions to those samples of the Martian atmosphere taken by the two Viking Lander probes and unlike those found on Earth or in other meteorites. The chemical composition of the meteorites was also found to be an exceptional match with Martian soils measured by the same space probes.

AURORAS

"The aurora is among the most beautiful and spectacular of all natural phenomena. No artist's brush, no camera can hope to capture the elusive nature of its constantly changing form, the purity of its colours or the complexity of its spatial dimensions. The aurora has to be seen to be believed. There are few verbal descriptions that do it justice. Only in the folklore of primitive peoples and sometimes in literature does one find true poetic insight. The Maoris of New Zealand describe the southern lights as 'Tahu-Nui-A-Rangi' – the great burning of the sky." (H. G. R. King)

Antarctica provides the ideal site for the study of complex processes present in the upper atmosphere, whose most prominent manifestation is the aurora australis or "southern dawn". Scientific observations from a number of bases, combined with imagery and information obtained by balloons, rockets and satellites, have given us a much better understanding of these wonderful yet mysterious phenomena.

A sea of plasma
First, it is necessary to understand that the planets are, in reality, frozen grains of matter floating in a sea of plasma – ionized gases – that make up 99.9 per cent of the universe. (Plasma is derived from a Greek word meaning to form round something.) There are many kinds of plasma, with different energies and chemical compositions, which maintain discrete forms even if they occur in the same places.

Surrounding the Earth, at the edge of the atmosphere, plasma is held and channelled by the Earth's magnetic field in a region called the magnetosphere. This is compressed on one side of the planet and drawn out into a huge comet-like tail on the other side by the force of the solar wind, a supersonic stream of plasma which flows away from the sun at a speed of several hundred kilometres per second.

At the North and South Poles, there is a gap between the magnetic lines of force on the day and night sides of the planet. At these polar cusps plasma from the solar wind penetrates down magnetic field lines and into the polar upper atmosphere.

A giant TV set
It is this bombardment that generates auroras, as the incoming electrons agitate the charged particles in the ionosphere and collide with the atoms and molecules in the electrically neutral thermosphere below it, causing them to emit photons. The colours of the auroras depend on the thermosphere's constitution – violet for nitrogen, red and green for atomic oxygen.

RARE AURORA
This false-colour satellite image from the spacecraft, Dynamic Explorer I, shows a "theta" aurora in which the auroral oval is bisected by an arc that takes 12 hours to spread across the polar ice cap. This occurs when the magnetic field is directed north.

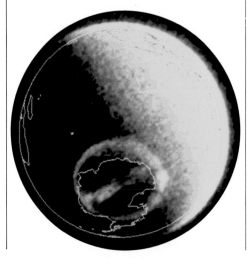

The best analogy is to view the entire area as a giant television set. The plasma tail corresponds to the electron gun, the polar atmosphere is the screen of the picture tube, and the aurora is the image you see on the screen.

Magnetic storms
The interaction of the solar wind with the Earth's magnetic field is a process subject to periodic fluctuations and disturbances. A major solar flare produces dramatic disturbances in the magnetosphere, known as magnetic storms, and at these times auroral activity is most intense and occurs at lower latitudes. The energy from the solar wind intensifies from 10,000 megawatts to 15 million megawatts during such periods. Such disturbances also disrupt shortwave radio communications.

Pollution
Antarctic scientists have, in the last two decades, discovered that electromagnetic radiation from the power lines of the industrialized world is modifying processes in the magnetosphere. Several Antarctic stations, formerly electromagnetically "quiet", have been receiving signals that arise from this little-known form of pollution.

Ground-based observations from Antarctica will play a major part in the International Solar Terrestrial Physics Programme (ISTP), planned for the early 1990s – a large-scale, coordinated campaign of scientific research involving simultaneous measurements from many points in the geospace.

AURORA AUSTRALIS OVER HALLEY BASE
Situated on a polar cusp, the British Antarctic Survey's Halley Base is ideally suited to observe the dramatic aurora australis. Since 1981, the base has operated a computer-controlled iosonde.

This radar device transmits pulses of radio waves at varying frequencies into the ionosphere and detects the reflected echoes in order to measure the electron concentration. It is these electrons that, colliding with the thermosphere, cause the aurora to form.

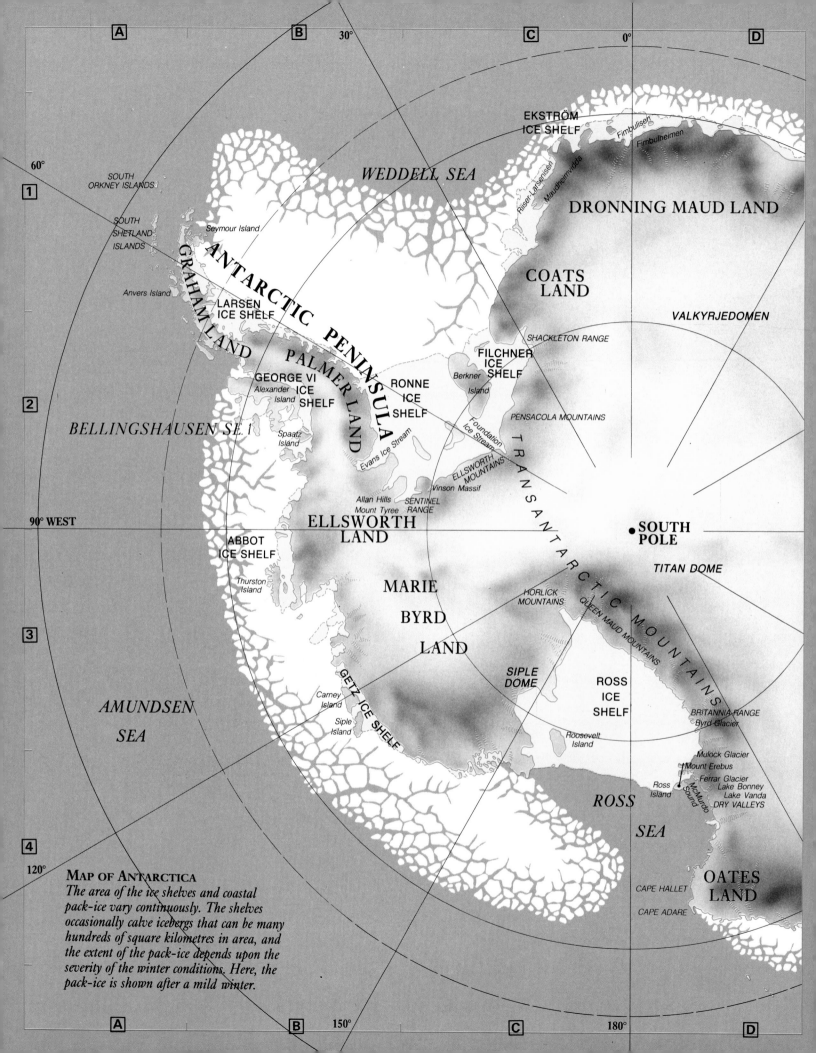

MAP OF ANTARCTICA
The area of the ice shelves and coastal pack-ice vary continuously. The shelves occasionally calve icebergs that can be many hundreds of square kilometres in area, and the extent of the pack-ice depends upon the severity of the winter conditions. Here, the pack-ice is shown after a mild winter.

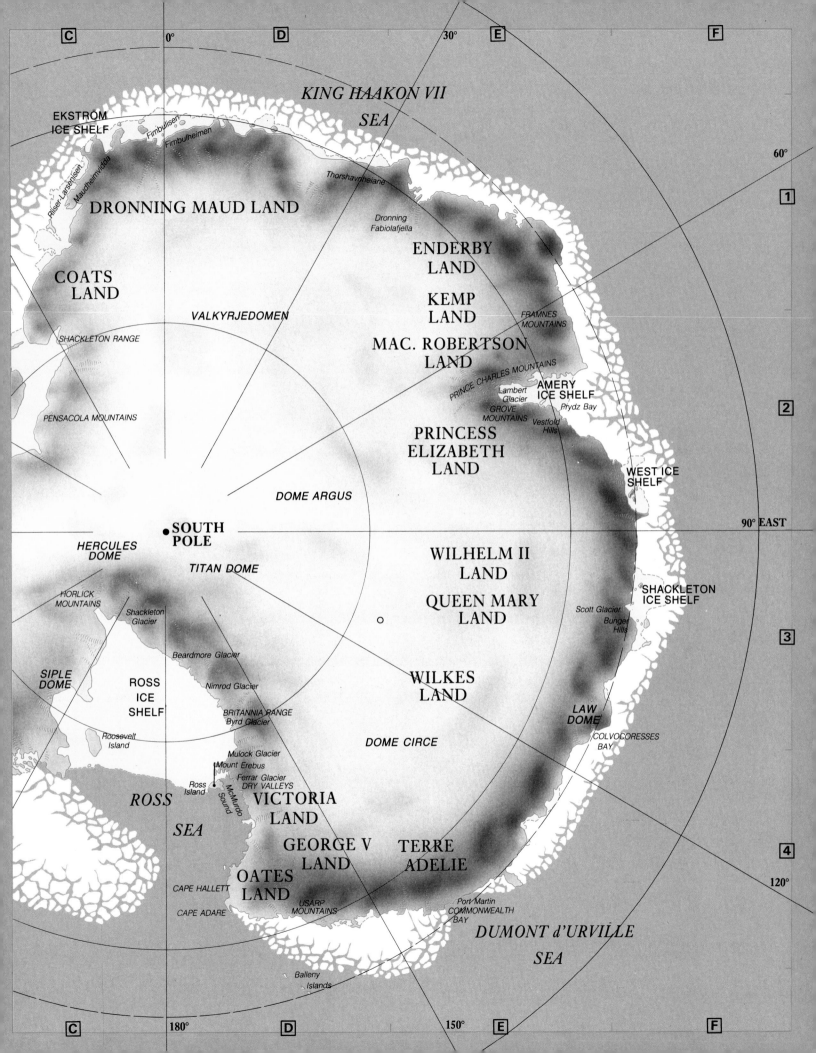

LIFE AT THE END OF THE WORLD

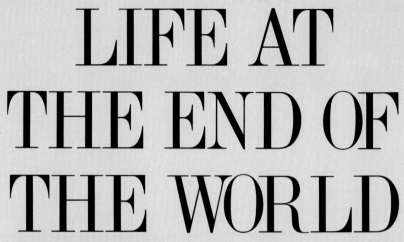

"No matter how we care to divide the phenomenon of life, regardless of the names we choose to give to species or the shapes we devise for family trees, the multifarious forms of life envelop our planet and, over aeons, gradually but profoundly change its surface. In a sense, life and earth become a unity, each working changes on the other."
Lynn Margulis and Karlene V. Schwartz –
Five Kingdoms

LAND FLORA & FAUNA

The harsh combination of rapidly changing temperatures, strong drying winds, irregular water and nutrient supply, frequent snow falls and frosts, and continual soil movement due to freezing and thawing, means that only a few plants and invertebrates can survive on the ice-free land (amounting to only two per cent of the continent) of Antarctica.

The areas capable of supporting life are the Dry Valleys in Victoria Land – cold deserts with saline lakes; small areas of the continent where the winter snow melts in summer, providing water for a variety of organisms; and the exposed rock faces on isolated mountains.

Survival techniques

The survival capacities of these highly specialized organisms are astonishing. Communities of bacteria, algae, fungi and small arthropods have been found in the most barren environments, including mountains close to the South Pole. Antarctic plants have survived by maintaining the vital processes of photosynthesis and respiration at temperatures which often fall below −10°C (14°F). The flora on the continent is dominated by lichens and mosses and, on a microscopic level, algae, fungi, and different forms of bacteria.

Most of the potentially suitable sites for vegetation are occupied by penguin rookeries, whose trampling and excrement kill many plants. Sheets of green algae can survive such conditions, however, and brightly pigmented algae also cause green, yellow and red snow on glaciers and ice caps. Microscopic filaments of fungi occur in the soil which, with bacteria, are responsible for breaking down dead plants to form peat and releasing nutrients into the ecosystem.

A proliferation of lichens

Lichens, of which more than 350 species are known, have proliferated in the Antarctic mainly because there is little competition from mosses or flowering plants and because of their high tolerance of drought and cold. Although lichens are able to colonize

RED SNOW

Green algae (Chlorophyta), *containing red pigments, form communities on the surface of the snow and ice (below). This "red snow" is a phenomenon of the snowfields near the coast and is most common in the Antarctic Peninsula. Capable of existing closer to the Pole than any other plants, lichens of varied colours, from grey and black to orange (right), cling to the exposed surfaces of windswept rocks, freezing in winter and recovering only slowly during the spring.*

soils which are poor in nutrients, they mainly grow on stones and rocks.

During the Antarctic winter, most lichens are inactive and their metabolism remains at a very low level. Some may freeze-dry, only to rehydrate when liquid water or water vapour becomes available. Some lichens can survive for at least 2,000 years.

Flowering plants

Only two flowering plants grow south of 60°S – the Antarctic hair grass (*Deschampsia antarctica*) and the Antarctic pearlwort (*Colobenthos subulatus*). They occur in small clumps near the shore of the west coast of the Antarctic Peninsula. Here, in the wetter areas, are found mosses and liverworts and on the drier, more exposed, sites – lichens. On the better drained, stony slopes mosses build up to a deep peat (as much as 2m/6ft 6in deep and 5,000 years old).

Microbes beneath the rocks

In the Dry Valleys, where hundreds of square kilometres of land are totally barren, it was realized recently that the most likely habitat to support life was not in the soil but in the rocks.

In 1978 biologists from Florida State University found a widespread rich microbial vegetation under the surface of some rocks. Dark green organisms, making a layer a few millimetres deep, were found in the air spaces of porous rocks or in fissures.

Algae, bacteria, and fungi were found to colonize light-coloured, semi-translucent rock, which the sunlight penetrates. The micro-climate between the minute grains of rock is warm enough to support life and the sunlight provides energy for photosynthesis. The uppermost rock layer protects the micro-organisms from drying up and from damage due to excessive radiation.

The organisms live on carbon dioxide found in the atmosphere, which filters through cracks in the rocks, on minute amounts of minerals in the soil covering the rocks, and on small traces of moisture that seep into the rocks from the small amounts of snow in the area. They switch their metabolic activity on and off, adapting to periods of sunlight and darkness. Some are believed to have been in the rocks for 200,000 years.

MOSS AND GRASS *Moss intermingles here with the only grass that thrives in Antarctica,* Deschampsia antarctica.

Microscopic invertebrates

The terrestrial fauna of Antarctica consists entirely of invertebrates, mostly microscopic, which live in the soil and in vegetation. They range from protozoa (single-celled creatures), rotifers, tardigrades and nematodes to arthropods (mainly mites and springtails). The largest invertebrate is the wingless midge *(Belgica antarctica)*, which grows to 12mm (½in) long.

Of the approximately 112 species of land arthropods in Antarctica, about 58 are free-living, 51 are parasitic and three are predators on other animals or semi-parasitic. The mites are the most widely occurring land animals.

Many of the arthropods avoid freezing by a physical process known as "supercooling", whereby their body fluids are maintained in a liquid state in temperatures below their normal freezing point. Species such as the oribatoid mite *(Alaskozetes antarcticus)* and the springtail *(Cryptopygus antarcticus)* have a constant struggle to maintain this unstable condition. The presence of food material in the gut provides some particles around which ice will form, so, in order to survive, they must strike a balance between freezing and starving. Their ability to synthesize glycerol, an antifreeze, enables them to survive temperatures of −35°C (−30°F).

Some of the densest populations of mites, springtails, nematodes and others – more than 1,000,000 per m² (100,000 per ft²) – have been found among algae growing on feathers of long-dead penguins and in mosses and algae in areas enriched by the debris of penguin rookeries.

WINGLESS INSECTS

The Antarctic springtail (Cryptopygus antarcticus), *seen here under the scanning electron microscope (above), is a wingless insect, 1mm in length, found mainly in the coastal habitats of the Antarctic Peninsula. It feeds on micro-fungi and algae.*

TERRESTRIAL MITES

The predatory mite, Gamasellus racovitzai, *attacks an antarctic springtail, its main source of food (right). This is the largest terrestrial predator in Antarctica, yet it weighs only 100μg. The majority of Antarctic mites, which live mainly in soil and vegetation, are barely visible to the naked eye.*

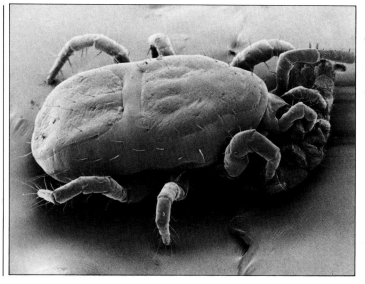

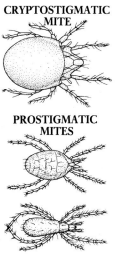

CRYPTOSTIGMATIC MITE

PROSTIGMATIC MITES

THE OCEAN FOOD WEB

The nutrients, plants and animals in an ecological system form an interdependent network. In the sea, organic matter and mineral nutrients are taken up by the microfauna and flora on which higher fauna feed. High animals feed on the fauna.

Antarctic terrestrial and fresh-water ecosystems tend to be relatively simple. Wide fluctuations in climatic conditions throughout the year militate against the successful establishment of a wide diversity of plant and animal forms.

The marine environment, on the other hand, is far richer and more stable. Conditions have remained virtually unaltered over thousands of years, and ocean currents have brought in a constant flow of new species, gradually building up a diverse web of interacting plants and animals.

Levels of life

The free-floating phytoplankton are the main source of primary production, blooming in Antarctic waters each summer. They are linked to the higher animal species through the zooplankton and demersal (bottom-living) fish, and through krill, squid and pelagic (upper water) fish.

Krill occupy a central position in this entire food web. It is clear that any threat to this species, through the kind of "over-fishing" that has so drastically reduced whale, seal and fish populations, would have repercussions throughout the ecosystem.

One particular chain in this network (phytoplankton-krill-whales) is of great significance because it is unusually short and therefore very fragile and susceptible to disruption. The survival of the great whales may depend on our gaining a better scientific understanding of this inter-relationship.

THE ANTARCTIC MARINE FOOD WEB

The inter-relationships of the plant and animal species in Antarctic waters, and the paths by which energy moves through the system, can be represented diagramatically in a food web (right).

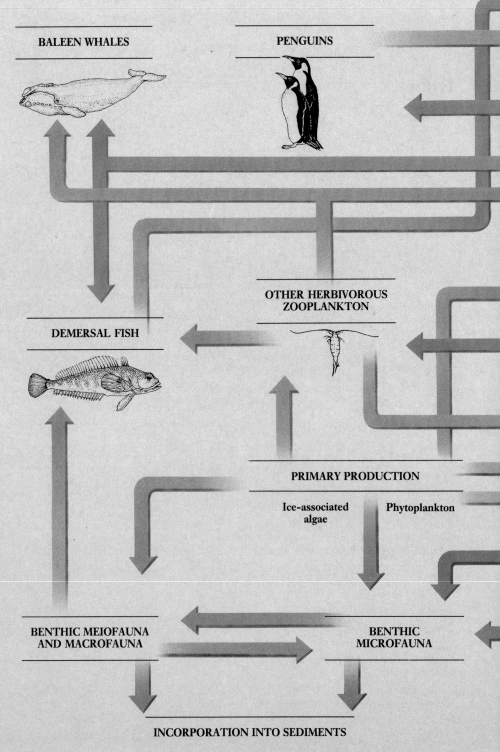

BALEEN WHALES

PENGUINS

OTHER HERBIVOROUS ZOOPLANKTON

DEMERSAL FISH

PRIMARY PRODUCTION

Ice-associated algae   Phytoplankton

BENTHIC MEIOFAUNA AND MACROFAUNA

BENTHIC MICROFAUNA

INCORPORATION INTO SEDIMENTS

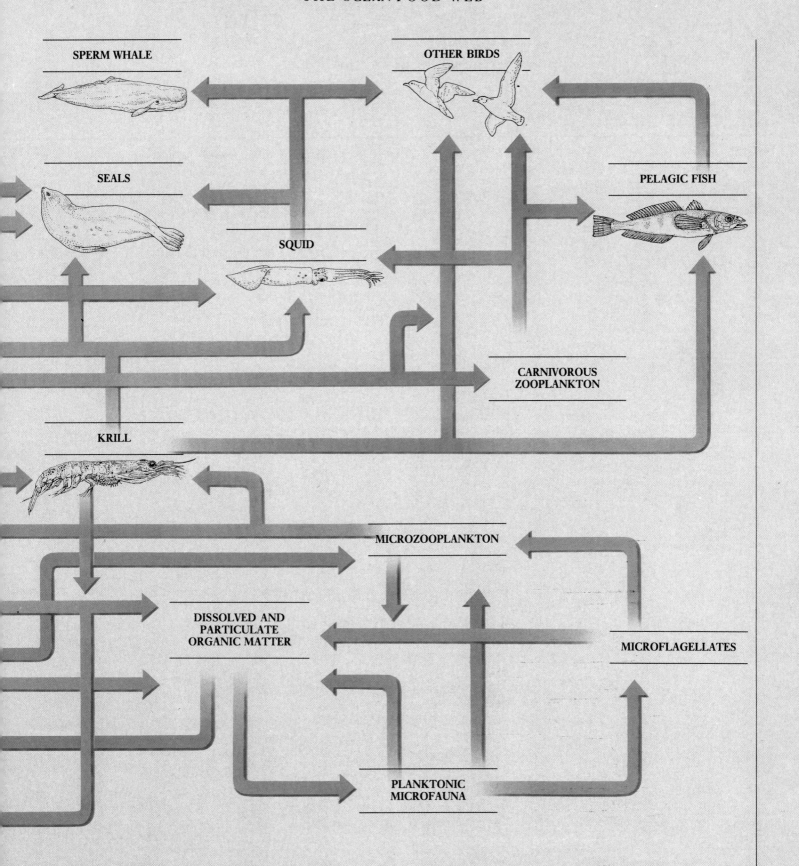

SPERM WHALE

OTHER BIRDS

SEALS

PELAGIC FISH

SQUID

CARNIVOROUS
ZOOPLANKTON

KRILL

MICROZOOPLANKTON

DISSOLVED AND
PARTICULATE
ORGANIC MATTER

MICROFLAGELLATES

PLANKTONIC
MICROFAUNA

FLORA & FAUNA OF THE SEABED

Most of the Antarctic coastline is hidden beneath the ice flowing off the continent. Beneath the seemingly lifeless expanses of pack-ice and sea-ice, however, lies the rich and varied world of the benthos – the flora and fauna living on the seabed around Antarctica.

There are some exposed coastal areas on the west coast of the Antarctic Peninsula and on the islands of the Scotia Arc, but even here, many areas are so scoured by ice that few organisms can survive. There are some seaweeds and large numbers of limpets but virtually no mussels or barnacles. In contrast, the seabed is teeming with life.

The limits of the benthos
The boundary of the benthos is the northern limit of the pack-ice rather than the Antarctic Convergence (the biological boundary for the sea creatures of the surface and upper waters). The boundary is determined by sediments carried from the land by icebergs and deposited when they melt.

An estimated 500 million tonnes of material are transported to the sea in this way every year, which is the equivalent of depositing 135 tonnes on each km^2 of the sea floor (400 tonnes per mile^2). These deposits range from silts and sands to boulders 50cm (20in) in diameter. The minerals in them dissolve in the water. In combination with the high concentration of oxygen in the cold water and the long daylight hours in summer, this creates a very rich medium for the benthic life-forms.

Complex communities
Diverse and complex communities of flora and fauna flourish at depths where the ice has little disruptive effect.

There are extensive beds of seaweed and luxuriant growths of other algae, which provide food and shelter for many animals, the most numerous being amphipod crustaceans. Small sea spiders feed here on hydroids and anemones. Nudibranchs (sea slugs) crawl over the rocky seabed. The ecological niche occupied by crabs in other ocean environments is here taken up by the giant isopod (*Glyptonotus antarcticus*). Starfish and sea urchins are also common. Corals, sea cucumbers, sponges and anemones decorate every exposed surface on the rock face.

Life at the greatest depths
At depths below 30m (100ft), conditions are very similar around the whole continent. Here the fauna is dominated by filter-feeders. Sponges, anemones and corals are anchored to the seabed, while molluscs, crustacea and starfish move over them.

In these regions, temperatures are low and stable, and food is scarce throughout most of the year. As a result, animals function at low metabolic rates, growing larger and more slowly than their counterparts in warmer waters and, in many cases, living longer. Some species of limpet can survive for more than 100 years; some sponges for several centuries.

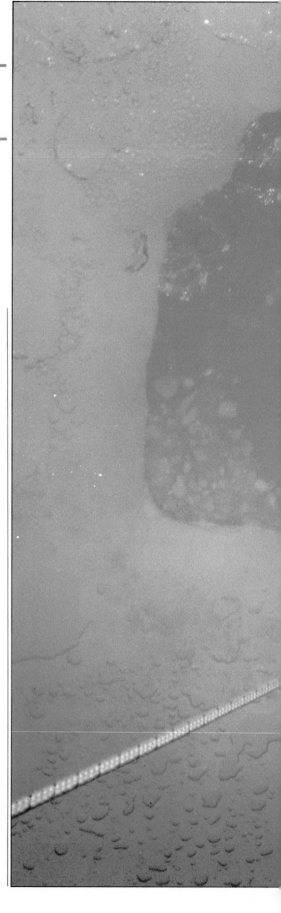

LIFELINE *A scientist returns from a dive beneath the ice off Signy Island, South Orkneys.*

LIFE ON THE SEABED

In contrast to the limited diversity of the terrestrial fauna, Antarctic benthic life forms are extremely varied and form the focus of much biological research.

Here on the seabed, the ice is a major factor, even at depths below 30m (100ft). In its various forms the ice affects water temperature and salinity, light levels, currents and the sediments which enrich the sea with dissolved minerals.

The extent and duration of the winter ice cover, and the damage that it causes vary widely from year to year.

The beaches between the tide lines are left bare by the abrasion of the ice. The fast-ice and pack-ice abrade the seabed down to 15m (50ft), and many species only move into shallower waters during the austral summer.

On some parts of the coast, thin ice platelets form on the seabed, encasing plants and animals and eventually tearing them away.

Icebergs driven onshore can bulldoze the sediments, causing damage to areas of the benthos, which may take years to re-establish themselves.

CRUSTACEA

The crustacean, Glyptonotus antarcticus (below), an omnivorous isopod some 12cm (5in) long, is to be found all around Antarctica. Close to the surface, the 5cm-(2in-) long Antarcturus isopod (below right) walks among the fronds of algae. In deeper waters the giant pycnogonid sea spider (bottom), with a leg span of 25cm (10in), picks its way across the seabed.

DIVERSE SEA FAUNA

On a sheltered overhang, 5m (15ft) below the surface of Signy Island, a multi-coloured world of anenomes, long hanging stalks of soft coral, bryozoans (sea mosses) and sponges, covers the rock surface (right).

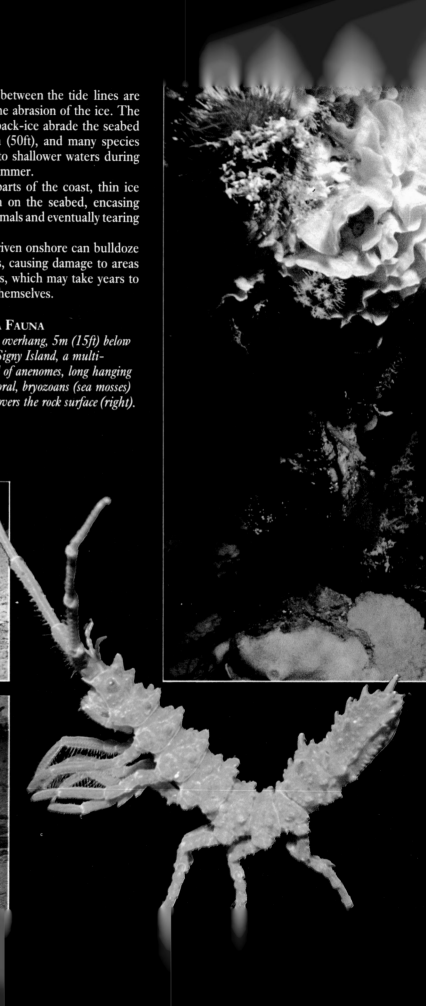

MOLLUSCS

Seen here spawning in the waters off Signy Island, the limpet, Nacella concinna, *is a common sight on much of the Antarctic shoreline, reaching densities of 350 per m² (35 per ft²). While the tide is up, it grazes freely, and when the water recedes it clings to the rocks (left).*

As winter approaches, large numbers migrate to deeper water in order to avoid being damaged by the scouring ice along the water's edge.

KRILL

Krill, a Norwegian whaling term meaning "small fry", refers to many species of planktonic crustacea. However, the dominant species is a semi-transparent shrimp-like creature, called Euphausia superba, *the Antarctic krill. It is the largest of the genus, with an average body length of 4cm (1½in), and a lifespan of up to seven years.*

Krill are omnivores; they use a specialized arrangement of progressively finer filters to extract a wide variety of micro-organisms and other crustacea from the water. They are also cannibalistic. In the summer they feed in the surface layers of the ocean, descending to deeper water during the winter, where they feed on plankton detritus.

Life cycles

The main spawning season for krill is January to March. Their eggs are released near the water surface and, over a period of about 10 days, sink into deeper water, developing as they go. The currents carry them south to the edge of the continental landmass where they hatch. The larvae gradually rise again to the surface, passing through two developmental stages before eventually emerging at the surface as juvenile krill.

In the late 1970s two American researchers discovered that krill exhibit a quality called "regression"; that is, they become smaller and less sexually mature in appearance after spawning. What had previously been taken to be immature females were in fact adults that had regressed; the krill population has a lower reproductive potential than previously thought. Many population studies and ideas about krill life cycles have had to be rewritten as a result of these findings.

A major source of food

The Antarctic Convergence provides a natural barrier to the krill, which provide the major food supply for five species of whale, three species of seal, 20 species of fish, three of squid and numerous bird species, including penguins. This dependence of so many predators on just one prey group is extremely unusual.

Although krill are circumpolar in distribution, there are several areas of high concentration which correspond to a series of eddies in the oceanic circulation. Huge swarms comprising more than 2 million tonnes of krill and covering an area of 450km^2 (170mls^2) have been observed. During the day they appear as giant red patches on the ocean; at night, their bioluminescent organs shimmer in the darkness.

A highly complex system between predators has evolved which permits all species to take advantage of this abundant food source. They feed in different areas, on krill of different ages, at different times of the year and at different depths. The huge swarms of krill allow even the largest predators, the baleen whales, to feed efficiently, but only during the summer when the ocean is not covered with pack-ice. Whale migration is adjusted to the krill life cycle.

ABUNDANT SOURCE OF FOOD

The main food source of many species, from fish and penguins to the great whales, the shrimp-like krill is the linchpin of the Antarctic ecosystem.

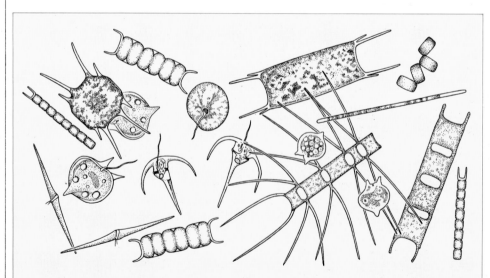

PREY AND PREDATOR

The primary food source for the krill is microscopic phytoplankton, the main plant producers of the ocean that "bloom" in the surface waters during spring and summer. The krill trap these abundant drifting plants with the fine hairs on their thoracic appendages. The krill in turn fall prey to the carnivorous squid, which form the main link with species higher up the food chain. The squid also feed on fish and other squid, drawing food into their horny jaws or "beaks" with powerful suckers on their tentacles.

Sperm whales and several species of bird eat an estimated 30 million tonnes of squid every year.

FISH

Of the 20,000 known species of fish in the world, only about 120 live in the waters of the Southern Ocean, south of the Antarctic Convergence.

A small proportion of these are deep-sea species of which little is known. The best studied group belongs to a single sub-order, the Notothenioidea, which account for more than 60 per cent of the coastal species and 90 per cent of the total number of individual fish. This sub-order, which comprises 84 species in the Antarctic, is divided into four families: Antarctic cod (Nototheniidae), plunder fish (Harpagiferidae), Dragonfish (Bathydraconidae) and ice fish (Channichthyidae).

"Supercool" fish

These polar fish have become cold-adapted over the last 40 million years, living in a "supercooled" state in which their body fluids remain liquid at temperatures below the point at which ice forms. They are able to maintain this state because their bodies contain up to eight different anti-freeze molecules (glycopeptides) which impede crystal growth and prevent ice from spreading through the body fluids. Exactly how this works is not yet fully understood.

Antarctic fishes have limited energy stores and must conserve energy, particularly in the winter. In order to avoid having constantly to resynthesize these vital anti-freeze molecules, the kidneys of these fish have cells that draw only selected wastes from the blood by a secretory process, leaving the anti-freezes in circulation.

The Antarctic toothfish

Notothenioids are primarily bottom-dwelling fish that do not have the gas-filled sacs, known as swim bladders, found in many other fish. However, the largest of the Notothenioidea, the Antarctic toothfish *(Dissostichus mawsoni)*, has evolved some specializations which make it neutrally buoyant and allow it to live efficiently in middle-level

waters. These features include a light-weight cartilage skeleton, hollow vertebrae and an abundance of fat, which is less dense than sea water.

The ice fish

The most unusual fish in Antarctic waters are the Channichthyidae, known as ice fish because of their white gills. They are the only vertebrates whose bodies entirely lack haemoglobin, the red oxygen-carrying pigment in the blood. As a result, their blood is almost translucent, with a yellowish tint.

As they have no haemoglobin, the oxygen-carrying capacity of these fish is only 10 per cent of other notothenioids

ANTARCTIC ICE FISH *The "white-blooded"* Chaenocephalus aceratus *(above).*

in the same environment. They compensate for this with a number of special adaptations suited to low temperatures.

Their blood volume is higher, their hearts are twice as large and they beat faster. However, they need less energy to circulate the blood because their blood vessels are larger and the viscosity of the blood is lower due to the absence of red blood cells. The transfer of oxygen from blood to tissues is more efficient and they use less energy to maintain their metabolism when resting than red-blooded fish.

ANTARCTIC COD *(Notothenia coriiceps)* a member of the "spine-rayed" fishes.

ANTARCTIC FISH

Antarctic fish species show remarkable physiological adaptations to their environment. Most species have predictably low metabolic rates, but some are "cold-adapted" and able to maintain a higher rate of protein synthesis than would be expected.

Some Antarctic cod display neutral buoyancy, achieved through changes to the skeleton and body lipid content, enabling them to remain in the midwater zone where krill abound. Others have anti-freeze in their blood. The ice fishes lack haemoglobin and their gills and internal organs remain colourless.

TOOTHFISH *(Dissotichus mawsoni)* a giant cod which reaches a length of 1.5m (5ft).

Channichthys rhinoceratus

Chaenocephalus aceratus

THE ICE FISHES

Pseudochaenichthys georgianus

SMALL ANTARCTIC HERRING
(Pleurogramma antarcticum)

MARBLE PLUNDER FISH
(Pogonophryne marmoratus), a species of Antarctic sculpin

GREEN ANTARCTIC COD
(Notothenia gibberifrons)

DEEP-WATER DRAGON FISH
(Bathydraco marri)

ANTARCTIC COD
(Trematomus bernacchii)

RAT-TAIL *(Lionurus filicauda)*, a deep-sea fish

EEL POUT *(Lycenchelys antarcticus)*, a coastal fish

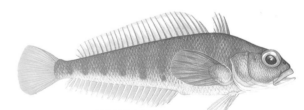

SEABIRDS

The popular myth is that the Southern Ocean is teeming with seabirds, and indeed nowhere is there a greater proportion of sea to land than in the southern hemisphere between latitudes 50° and 70°S. In fact, because the total area of land suitable for breeding seabirds is so restricted, the birds are clustered together.

The majority of the region's birds depend upon the 24 oceanic island groups as breeding grounds. Only 19 flying species breed on the Antarctic continent. Consequently, visitors to Antarctica receive a distorted picture of the relative abundance of birds over the whole region.

On a typical sub-Antarctic breeding island, penguins occupy the coastal slopes, cormorants and smaller albatrosses the cliffs, larger albatrosses the higher, flatter ground, and small petrels the coastal lowlands and inland slopes.

Eating habits

What seabirds eat and how they obtain their food is in many cases still a matter for some speculation. Land-based studies have ascertained the nature of the diet of breeding birds during the austral summer, but their diet and feeding habits while they are at sea is little known. What is known is that many species feed at night when their prey is on the surface of the sea. New research suggests that albatrosses and petrels use their sense of smell extensively to locate their food.

Seabirds show a variety of feeding methods at the surface layer, which extends to 3m (10ft). Storm petrels and terns feed by shallow dipping through the surface film; albatrosses, petrels and diving petrels use shallow surface dives or plunges.

By contrast, the small, smoky-blue coloured prions, or whale birds, filter-feed by paddling along the surface of the water with their blue bills immersed. Their bills contain tiny comb-like plates which trap food particles; the skin of their mandibles forms a pouch, which shoots water out from the sides.

In general terms, 55 per cent of the seabird population (excluding penguins) eat plankton; 20 per cent, mainly the albatrosses, eat squid; 20 per cent have a mixed diet of plankton and squid and a fourth group (5 per cent) eat fish.

Plankton eaters are, on average, much smaller than the squid eaters and tend to occur in the south of the Antarctic region. They feed more frequently than the larger birds, and on bigger concentrations of smaller prey.

Large birds have proportionally lower metabolic rates and take larger prey, which are more sparsely dispersed.

It is estimated that birds consume, either directly or indirectly (eg, squid eat krill), at least 100 million tonnes of krill annually, of which about 90 per cent is taken by penguins.

The struggle for survival

Southern Ocean seabirds are generally restricted to breeding during a short summer. All the procellariiform species (eg, the albatrosses) lay one egg and both parents share incubation, brooding and chick-feeding duties. Incubation is unusually long compared to that of any other birds – up to 80 days.

This adaptation is believed to have evolved owing to the difficulty parents have in finding not only the necessary quantity but also the quality of food for themselves and their chicks. Breeding success in many species is usually less than 50 per cent, due mainly to the starvation of the chicks. This high mortality rate is balanced by the high life expectancy of adult birds.

Owing to the long incubation periods, parents and chicks stay on land for extended periods. As a result, the birds are there long enough to support populations of avian scavengers (kelp gulls

SMOTHERED BY THE SNOW
A giant petrel (Macronectes giganteus), *half buried by a blizzard, keeps its egg warm beneath the snow.*

and sheathbills) and predators (skuas). The birds also modify the nature of the terrain by depositing organic matter – excreta, feathers, carcasses – and by burrowing into the ground.

PROCELLARIIFORMES

Most of the Antarctic's flying seabirds belong to two orders – coastal-based skuas, gulls and terns (Charadriiformes) and ocean-ranging albatrosses and petrels (Procellariiformes).

This order is also called the Tubinares because its members have obvious tubular nostrils on top of the beak or on each side of it. The nostrils have glands at the base which are used to excrete excess salt. The order is represented by members of all four of its families – six species of albatross (Diomedeidae); 23 species of petrel, prion and shearwater (Procellariidae); four species of storm petrel (Oceanitidae); and two species of diving petrel (Pelecanoidae). They range in

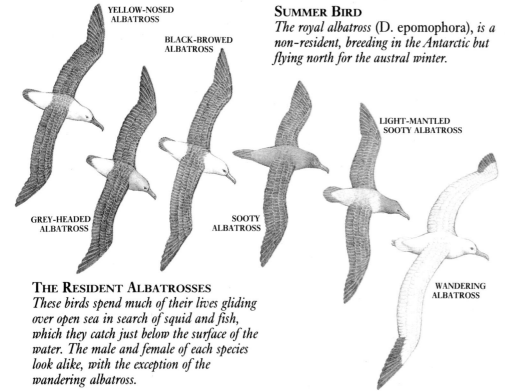

YELLOW-NOSED
ALBATROSS

BLACK-BROWED
ALBATROSS

GREY-HEADED
ALBATROSS

SOOTY
ALBATROSS

LIGHT-MANTLED
SOOTY ALBATROSS

WANDERING
ALBATROSS

SUMMER BIRD
The royal albatross (D. epomophora), *is a non-resident, breeding in the Antarctic but flying north for the austral winter.*

THE RESIDENT ALBATROSSES
These birds spend much of their lives gliding over open sea in search of squid and fish, which they catch just below the surface of the water. The male and female of each species look alike, with the exception of the wandering albatross.

size from the wandering albatross, which has a wingspan in excess of 3.3m (11ft) and weighs between 7.7 and 9.5kg (17 and 21lb), to the storm petrel, which weighs a mere 30 to 60g (1 to 2oz).

Albatrosses (Diomedeidae)
The six species of this family that are resident in the region are the wandering albatross *(Diomedea exulans)*, the black-browed albatross *(D. melanophris)*, the grey-headed albatross *(D. chrysostoma)*, the yellow-nosed albatross *(D. chlororhynchos)*, the sooty albatross *(Phoebetria fusca)* and the light-mantled sooty albatross *(P. palpebrata)*. The estimated breeding population of all six species is 750,000 pairs.

The wandering albatross is the biggest and is very similar to the royal albatross *(D. epomophora)*. They breed only once every two years and the chicks take nine months to fledge.

These birds mature quite late, starting to reproduce at 10 to 14 years, and

85

have a lifespan of 80 to 85 years. The wandering albatross has more wing feathers than any other bird.

Some of the smaller albatrosses are also known as mollymauks.

Petrels (Oceanitidae, Pelecanoidae, Procellariidae)

All petrels have dense plumage and webbed feet. Prions, gadfly petrels, shearwaters, storm petrels and diving 'petrels tend to nest on the sub-Antarctic islands in burrows in the soil or in cracks in the lava. They normally only visit land at night. Many of these species are small, look alike and live in rugged terrain, making an accurate census of their numbers difficult, but their estimated total population is 150 million birds.

Some species, such as the snow petrel *(Pagrodama nivea)* and the Antarctic petrel *(Fulmarus glacioides)*, can breed several hundred kilometres inland on isolated nunataks on the ice sheet. Most of the other petrels breed along the coasts, particularly the west coast of the Antarctic Peninusula and the Victoria Land coast on the Ross Sea.

SNOW PETRELS *Living in Antarctica all year, snow petrels feed mainly on fish.*

The giant petrels *(Macronectes giganteus* and *M. halli* – southern and northern species) are colloquially known as stinkers. When threatened, they project a stinking stream of oily vomit.

CHARADRIIFORMES

Cormorants, gulls, sheathbills, skuas and terns are coastal birds rather than deep water seabirds: they forage close to shore. There are three species of tern, one of gull (Laridae), two of skua (Stercorariidae), two of sheathbill (Chionididae) and two species of cormorant (Phalacrocoracidae).

Cormorants (Phalacrocoracidae)

The two species of cormorant – the blue-eyed shag *(Phalacrocopax atriceps)* and the king shag *(P. albiventer)* – are found on the Antarctica Peninsula and islands of the Scotia Arc, and on Kerguelen Island respectively. They breed close to the sea in colonies of a few dozen pairs, building nests out of seaweed and feathers held together with guano. They dive for fish, squid, benthic worms, molluscs and crustacea.

BLUE-EYED SHAGS *These birds (right) overwinter on Petermann Island.*

Gulls (Laridae)

The southern black-backed gull (*Larus dominicanus*) is widespread on the Antarctic Peninsula and on many of the sub-Antarctic islands. It lays two or three mottled eggs in November or December. These hatch in less than a month, and the chicks reach independence six to eight weeks later. A scavenger, it feeds along the shoreline.

VULNERABLE CHICK

A juvenile gull (above) hungrily awaits the return of its parent. Unable to fly, the chick risks falling prey to skuas.

EVENING FLIGHT

A Dominican gull flies in the late evening off the coast of the Antarctic Peninsula.

Sheathbills (Chionididae)

Small, plump, white, pigeon-like birds, sheathbills are the only birds in Antarctica without webbed feet. They breed in solitary pairs, laying two to four eggs in nests concealed in rock crevices.

The snowy sheathbill *(Chionis alba)* is found in the Antarctic Peninsula and islands of the Scotia Arc; and the black-faced sheathbill *(C. minor)* on other sub-Antarctic islands. Omnivorous birds, they will consume absolutely anything, from insects, plants and eggs to the carcasses of dead animals and even the excrement of seals.

Skuas (Stercorariidae)

The Antarctic skua *(Catharacta maccormicki)* and the brown skua *(C. lonnbergi)* are almost identical in appearance and behaviour. Notorious predators and very aggressive, they nest near penguin colonies, which they raid for eggs and young birds.

Outside the breeding season, they roam the oceans feeding on krill and small fish before returning to the same breeding grounds every year with the same partner. In winter they migrate to the North Pacific, keeping company with flocks of shearwaters.

They lay two eggs in a rough nest made of bones, pebbles and moss, but only rear the first chick to hatch.

Terns (Laridae)

The Antarctic tern *(Sterna vittata)* breeds on the Peninsula and islands in small colonies, feeding on small fish and plankton, which it finds just off the coast or in tidal pools. Its eggs and chicks are camouflaged to blend in perfectly with their environment.

The Kerguelen tern *(Sterna virgata)* is limited to islands north of the Convergence and feeds on a diet of small molluscs, and insects and their larvae.

These two species are supplemented during the summer months by large numbers of the Arctic tern *(Sterna paradisaea)*, which migrate more than 40,000km (25,000 mls) from the other end of the world to feed here before returning northwards. No other bird makes such a long journey annually; it means that, with the exception of a few weeks while it is on the wing, the Arctic tern spends much of its life in daylight.

SKUA STUDIES BY WILSON

Skuas are the most unattractive Antarctic birds both in appearance and character. Brown-bodied with flecks of white on their wings, they have ugly hooked beaks. They feed on penguin eggs and chicks in addition to adult seabirds, and will attack humans if they come too close to their nests.

THE PINTAIL AND THE PIPIT

South Georgia is home to the only two land birds (apart from penguins) that live south of the Antarctic Convergence all year round. The yellow-billed pintail *(Anas georgica)* nests in swampy areas in the shelter of big tussocks. The South Georgia pipit *(Anthus antarcticus)*, Antarctica's only true songbird, nests on small islands off the coast, to avoid the rats which have been accidentally introduced by man. It forages on the ground for insects and invertebrates.

PENGUINS

Of the 18 living species of these flightless birds, seven live within the Antarctic region. The auks were the first birds to which the name "penguin" was applied. They gained the name owing to the quantity of fat (penguigo) obtained from them.

The origins of penguins are obscure. They appear to have evolved from gull-like or petrel-like flying birds of the Eocene era (40 to 50 million years ago), and to have passed through a stage when they could both swim and fly. For most of their evolutionary history, penguins inhabited warm seas but they gradually adapted to colder temperatures during the long period when the Southern Ocean cooled.

They are now equipped with dense plumage, each waterproofed feather overlapping its neighbour, and trapping a layer of air next to the skin. The feathers come complete with an underfelt of woolly down and there is a thick layer of fat between the skin and the underlying muscles. The fat also serves as an energy store for the birds.

So well insulated are they that penguins can become overheated. If this happens they ruffle their feathers to allow the insulating air layer to escape. Bare patches on the face, flippers and feet also get rid of surplus heat.

Underwater birds

Ungainly on land, penguins are superb swimmers, "flying" through the underwater environment with ease and grace. Their partially webbed feet and triangular tail form a rudder, while their flippers, powered by huge wing muscles, provide the motive power to push them through the water at great speed. They "porpoise" in and out of the ocean, snatching breaths of air and coating themselves with a skin of air bubbles that reduces friction.

Though short-sighted in the air, penguin eyes are clearly focused underwater, their retinas especially sensitive to violet, blue and green light. Penguins have spiny tongues and powerful mandibles to grasp their slippery prey. Like many seabirds that absorb a large

SOUTHERN OCEAN PENGUINS
These are the seven species of penguin to be found in Antarctica.

ADELIE *(Pygoscelis adeliae)*
Most abundant and widely distributed species.

CHINSTRAP *(Pygoscelis antarctica)*
Smallest of the "brush-tailed" species.

GENTOO *(Pygoscelis papua)*
Probably the fastest swimmer of all birds.

ROCKHOPPER *(Eudyptes crestatus)*
Most aggressive of all penguin species.

MACARONI *(Eudyptes chrysolophus)*
Confined to the outer and inner Antarctic islands.

quantity of salt either through their food or by drinking sea water, they have special glands that extract salt from the blood and excrete it through the nose.

Penguins have few parasites but several predators. Some species of bird, most notably the skua, scavenge their nesting sites; in the ocean, they are prey to leopard seals, sea lions and killer whales. Human disturbance has caused the most disruption, however. All penguin eggs are edible and have red yolks, which are derived from the pigment in the crustaceans they eat.

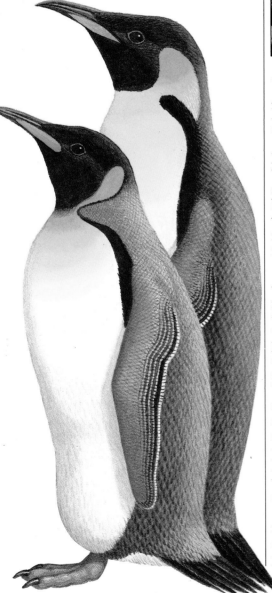

EMPEROR *(Aptenodytes forsteri)*
Largest and heaviest of all penguins.

KING *(Aptenodytes patagonica)*
Produce, on average, one offspring every two years.

There are three genera of penguin in Antarctica: Pygoscelis (brush-tailed), Endyptes (crested) and Aptenodytes (king and emperor).

PYGOSCELID PENGUINS
The three species of this genus are similar in size, shape and general behaviour. They are mainly black and white in coloration, and derive their name from their prominent tail of stiff feathers, which sweep the ground.

Adélie penguin *(Pygoscelis adeliae)*
Average weight: 5kg (11lb); average height: 70cm (27.5in). The most abundant and widely distributed of all penguin species, Adélies live and breed on the shores of continental Antarctica and on many of the Antarctic islands. They winter on the pack-ice, where the air temperature is higher than on land and in October begin moving south, often travelling 80km (50mls) or more, to their breeding sites. These are located on rocky slopes exposed to the wind, which means they are not smothered by snowdrifts.

The males arrive first and occupy the nest spaces. A few days later the females arrive. After courtship, they build their pebble nests and, in November, the two eggs are laid a few days apart. The female then goes to the ocean to feed on a diet of krill and larval fish. The male incubates the egg for seven to 10

NESTING ADELIES
These penguins incubate their eggs in nests of pebbles on the rocky shore, just out of pecking range of each other. The Adélie is the smallest of the true Antarctic penguins.

days, after which the female returns for a similar shift. The partners then alternate duties for the rest of the 35-day incubation period.

When the chicks hatch, they are warmed by the parents for two to three weeks before forming crèches. In February, the young birds leave the colony, followed a month later by the adults after they have moulted. From April on, the colonies are deserted. The young birds spend their first two years at sea or on the pack-ice.

The huge size of some Adélie colonies implies an equally huge food input. One biologist has estimated that the five million Adélies on the South Orkneys require 9,000 tonnes of krill and larval fish per day to feed their young at the height of the breeding season. This is the equivalent of the catch made by 70 trawlers, needing port facilities 10 times greater than those of Aberdeen.

Investigations have shown that Adélie penguins have remarkable homing abilities, and appear to navigate by the sun. In order to do this successfully, they probably have an inbuilt "biological clock", which takes account of the sun's changing position in the sky.

Chinstrap penguin
(Pygoscelis antarctica)
Average weight: 4.5kg (10lb); average height: 68cm (27in). The smallest of the three species, and with the narrowest geographical range, chinstraps occur mostly on South Orkney, South Shetland and the South Sandwich Islands, and along the western shores of the Antarctic Peninsula. Their total population may exceed 10 million and is expanding rapidly, according to some.

Chinstraps nest on rocky, sloping ground. Noisy and aggressive, they often infiltrate colonies of Adélies and take over their nesting sites. Females take the first incubation watch over two eggs and parental duties are shared once the chicks hatch.

Chinstraps feed almost exclusively on krill which they catch closer inshore than the other penguin species. Young birds leave the colonies within nine weeks, and all the birds abandon the shore by April or May at the latest.

Gentoo penguin *(Pygoscelis papua)*
Average weight: southern gentoo – 5.5kg (12lb); northern gentoo – 6.2kg (13.6lb); average height: southern gentoo – 71cm (27in); northern gentoo – 81cm (30in). Spread over a wide geographical range on the Antarctic and sub-Antarctic islands, gentoos have a total population of some 350,000; two-thirds live on South Georgia, with a few thousand in the far south.

Of the two subspecies, the southern gentoos *(Pygoscelis papua papua)* are smaller, lighter and have shorter feet, flippers and bills than the northern gentoos *(Pygoscelis papua ellsworthii)*.

Gentoos establish themselves on mounds amongst the tussock grass, and collect stones and scraps of vegetation to make their nests. They lay two almost spherical eggs the size of tennis balls, which have rough, bluish-white shells. These are incubated by the male and female in turn for 35 days.

Four or five weeks after hatching, the young chicks form crèches, which offer some protection from brown skuas, and both parents then hunt for food.

PROTECTING ITS YOUNG *A chinstrap penguin, with its distinctive facial markings, guards its family.*

Gentoos are probably the fastest swimmers of all birds. Estimates vary but they appear to be capable of bursts of speed of up to 21 to 27km per hour (13 to 17mph). They feed mainly on krill and other fish, and can dive to depths of 100m (328ft) or more.

UNDERWATER FLIER
The gentoo hunts for fish and squid by "flying" powerfully underwater (right). It builds a nest of seaweed and pebbles (below).

EUDYPTID PENGUINS
The penguins of this genus are of medium size, with rather stout bills and yellow crests on the sides of their heads.

Macaroni penguin
(Eudyptes chrysolophus)
Average weight: 4.2kg (9lb); average height: 70cm (27.5in). Confined to the outer and inner Antarctic islands, macaronis are found in their millions on Heard Island and South Georgia. They spend the winter at sea, returning to their colonies to engage in lively courtship before building their sparse nests.

Macaronis lay two eggs, the first being 50 per cent smaller than the second. This first egg rarely survives and is usually lost from the nest early on.

After a 35- to 37-day incubation period, the surviving chick is guarded by the male for the first three weeks while the female periodically returns with krill. From mid-January onwards both parents hunt for food and the young leave the colony by March.

Because of their bright yellow crests these penguins were dubbed macaronis, after the "Macaroni Dandies". These were 18th-century English travellers who adopted the more flamboyant European fashions.

Rockhopper penguin
(Eudyptes crestatus)
Average weight: 2.5kg (5.5lb); average height: 55cm (21.6in). By far the smallest of the Antarctic penguins,

93

ROCKHOPPERS *The female feeds the chick while the male guards.*

rockhoppers breed on the cliffs and scree slopes of islands close to and north of the Antarctic Convergence.

Found most often in the company of macaronis, they follow a similar breeding routine. They lay two eggs of different sizes; some pairs incubate both eggs but usually only one survives. Chicks leave the colonies when they are around 70 days old; they become the most aggressive of all penguins.

APTENODYTID PENGUINS
This wholly Antarctic genus comprises the two largest species of penguin. Their bills are long and slender, and curve gently downwards.

Emperor penguin
(Aptenodytes forsteri)
Average weight: 30kg (66lb); average height: 1.15m (3.8ft). The largest and heaviest of all penguins, emperors breed in some 30 known colonies on the Antarctic continent, all but two of which are situated on winter fast-ice, which is stable from late autumn onwards but

begins to break up in the spring. They are perhaps the only species of bird that never sets foot on land.

Virtually all our knowledge of these extraordinary creatures has been acquired in the last 30 years, although the first recorded sighting was on 12 October 1902 by Lt. Skelton of Scott's first expedition. New colonies were still being discovered as late as 1986.

By the time most other penguin species have completed their breeding cycles, emperors are only just starting theirs. The colonies assemble in April and May, at the end of summer, as soon as the new sea-ice is strong enough to hold their weight, at sites in the shelter of ice cliffs. Courtship takes three to five weeks: the partners identify each other with trumpeting calls.

The single eggs, which are 13cm (5in) long and weigh around 0.4kg (1lb), are laid in May and early June. They are the smallest eggs laid by any bird in relation to its body weight – 1.47 per cent of the adult body mass.

The male takes charge of the precious egg within a few hours, balancing it carefully on his feet and covering it with a special fold of abdominal skin.

The females return to the sea shortly afterwards and do not return until July.

For the first 65 days, the male emperors incubate their eggs. In the worst weather they mass together in sleepy, torpid huddles to reduce their exposed surface area. Conditions are extreme: the average temperature is around −20°C (−4°F) with winds occasionally gusting up to 200km per hour (124mph). They will have lost 40 per cent of their body weight by the time courtship and incubation are over, due to more than three months of fasting in the middle of winter.

In such conditions, any strategy to reduce heat loss is vital. Emperors even recover 80 per cent of the heat in their breath by an elaborate heat exchange system in their nasal passages.

The return of the females coincides with hatching. (Those chicks that hatch early are sustained by a secretion of protein and fat from the wall of the male's gullet.) The partners locate each other by calling, and the female then takes over the young chick. To replenish their reserves, the males leave for the ocean which, at that time of year, may be several days' journey away across wide stretches of ice.

For the next three to five weeks the chick grows slowly on food provided by the female. After that, both parents take turns to feed it. At around six weeks, the chicks gather in crèches and both parents search for food.

If the ice has begun to break up, allowing the parents easy access to the sea, the chick is fed regularly and prospers. If the ice has remained fast and the parents have to travel long distances, many chicks die of starvation. The majority do die; only four out of 10 survive to maturity.

By December or January, the chicks have begun to moult and, as the sea-ice breaks up, the colonies also begin to disperse. The young birds, still only half the weight of their parents, float out to sea on floes to complete their growth on the pack-ice. The adults moult in January and February, on floating ice close to the shore, then leave to spend the summer feeding at sea before starting the whole breeding cycle again.

Surprising as it is, winter breeding in the coldest place on Earth makes good sense for the emperor. If their chicks, which have a large food requirement, were reared in the summer, they would be left to fend for themselves in the autumn when food is low. This way the chicks gain independence at a time of maximum food supply. This option is not open to other Antarctic species. Only the adult emperor is large enough to survive the necessary winter fast.

King penguin *(Aptenodytes patagonica)*
Average weight: 15kg (33lb); average height: 95cm (3.1ft). Based on the Antarctic and sub-Antarctic islands, king penguins breed in huge colonies, which are situated on slopes close to an accessible beach. The colonies are occupied the whole year round either by chicks or adults. Their world population is estimated at two million.

Like the emperor, king penguins lay only one egg at a time, which they incubate by carrying it around on their feet, covering it with their brood patch – a fold of skin that covers the egg and transfers heat from the adult body – rather than in a nest.

They have two peak periods of egg laying. The "early breeders" lay their eggs in November, after which the parents take 15-day incubation shifts until the egg hatches from mid-January onwards. To feed their young, king penguins take fishing trips of four to eight days during which they will make hundreds of dives. Fewer than one in 10 of these dives ends in the capture of food. Nevertheless the chicks are raised to 90 per cent of their adult weight by April on a diet of squid and fish.

The "late breeders" lay and incubate from January to March. The earliest of these chicks can also reach near-maturity by the end of April. From then on, throughout the winter months, parental visits become less frequent and the chicks are forced to survive the blizzards and icy conditions on their own fat reserves, huddled together in their crèches.

When food resources increase in October, the parents return more often. As a result, the lean chicks fatten again and, by December, the largest of them have moulted and left for the sea. The parents of these survivors will then also moult and, in their new plumage, leave for the ocean for several weeks, returning to become the "late breeders" of the next season.

By this time, those penguins who have lost their eggs or chicks during the winter are already laying or incubating, forming the "early breeders" of that particular year.

Their unusual and lengthy breeding cycle, rivalled only by that of the Californian condor and the royal albatross, means that they can produce, on average, only one offspring every two years or, at most, two offspring during the course of three years.

By using depth recorders, scientists have estimated that of 500 to 1,200 dives made by king penguins, about half are greater than 50m (160ft). The deepest recorded dive for this species of penguin is 250m (820ft), twice the height of St. Paul's Cathedral. Only the emperor penguin dives deeper.

PROFILE OF A KING
Unlike the emperor, which breeds only on the Antarctic continent, the king penguin breeds on the islands around Antarctica, especially on South Georgia, Kerguelen and Macquarie.

A Year in the Life of the Emperor Penguin

Enabled by a greater body-weight to fast in the harsh conditions, only the emperor penguins hatch their eggs during the winter months. The males incubate them, grouping together to keep warm, each holding an egg on its feet for more than 60 days. When the chick hatches it is fed from the male's crop, until the female returns from the ocean to care for the chick, enabling the hungry male to go in search of food. Then both parents feed the young which becomes independent by mid-summer, when food is abundant.

In April the adult emperors "toboggan" in from the sea to the rookery.

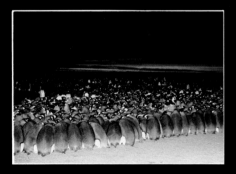

After the single egg is laid, the males huddle together against the winter cold.

For the next two months the males incubate the eggs on their feet.

It is late winter and bitterly cold. A two-week old chick snuggles into its mother's down to keep warm (above and top).

The month-old chicks stay in crèches while the parents seek food.

The well-fed chicks grow quickly after the fifth or sixth week.

By December they are ready to moult and, although they are not yet fully grown, they will soon leave their parents and fend for themselves.

SEALS

The Southern Ocean is an ideal habitat for seals, as they are extremely well insulated against the cold. Some have a layer of dense fur and others a layer of blubber beneath the skin, which is equally effective in the air or under water.

The scientific name for seal is pinniped, which means "fin-footed", and it is this characteristic that makes them ideally designed for an amphibious existence.

Seals are divided into two main types: the phocids or true seals, which evolved from otter-like ancestors, and the otariids, which include the fur seals.

PHOCID SEALS

The phocids make up the majority of Antarctic seals. They have no external ear, although their hearing is acute and equally good in air and in water. On land, they crawl or hump their way along in an ungainly fashion using only their short, clawed fore-flippers.

Crabeater seal
(Lobodon carcinophagus)
With a world population estimated at 40 million, this seal is one of the world's most abundant large mammals.

They are creatures of the drifting pack-ice, great wanderers throughout the Antarctic region but rarely seen ashore. Occasionally, some do make unexplained excursions inland; in 1957/58 two American scientists discovered the mummified carcasses of 90 of these seals at an altitude of 30 to 60m (100 to 200ft) in the Taylor Valley.

Despite its name, 90 per cent of the crabeater's diet is krill, supplemented by small amounts of other fish and squid. It has remarkable five-pointed teeth, the upper and lower rows of which interlock to form a strainer, so they can retain the krill while allowing water to be expelled. Crabeater seals in the Antarctic region are estimated to

SWIMMING FOR FOOD *Crabeater seals search for food beneath the pack-ice.*

consume 63 million tonnes of krill a year, which is more than the amount eaten by baleen whales.

Females are up to 3m (10ft) long and weigh up to 300kg (660lb); males are slightly smaller. Pups are born on the pack-ice in the spring (September to November). The mother suckles the pup for four weeks, during which it grows rapidly from 20 to 113kg (44 to 250lb) in weight.

During this period, the male rigorously defends an area of 50m (160ft) all around them, and there are frequent competitive fights with other males. Even the females are aggressive during this time.

Elephant Seal (southern)
(Mirounga leonina)
Largest of all seals and one of the largest of all mammals, fully grown male elephant seals can reach a length of 4.5m (15ft) and a weight of 4,000kg (8,800lb). By contrast, the females reach a maximum of 2.8m (9ft) and a weight of 900kg (2,000lb).

Covered in a uniform brown coat of short, stiff hairs, elephant seals get their name partly from their size, and partly from the male's trunk-like proboscis. Adults are frequently scarred: the males bear scars around the chest and neck due to wounds inflicted by other males; the females are scarred on the back of the neck, where the bull holds them during copulation.

With an estimated total population of more than half a million, elephant seals are established into three main breeding stocks, which are centred on a number of the sub-Antarctic islands. Their diet is mainly squid and some fish.

After spending much of their winter at sea, the first pregnant females come ashore in September on to the same sand and shingle beaches that they favour year after year. Some males have already arrived to stake out their territories and the cows form themselves into groups of 20 to 30 around each bull.

Males defend their harems vigorously, seeing off their rivals with deep, grating roars, their inflated trunks

acting like echo chambers. If this is not enough to scare them off the bulls will rear up and, occasionally, come to blows, delivering massive swipes to their rivals with their heads and bodies, slashing at them with their canines.

In this way a hierarchy is established. The most dominant bulls – the beachmasters – positioned in the middle of a group of cows with the subordinate bulls around the outside.

The young are born with coats of jet black woolly hair which they shed at three to four weeks. They weigh 46kg (100lb) at birth and double their weight in 11 days on a diet of rich milk.

Neither females nor males feed while ashore. Females fast for 30 days, and males for up to 90 in the case of beachmasters. Their role is so exhausting that few bulls are dominant for more than two seasons, during which they will breed with as many as 80 cows a season.

Throughout mid-summer the adults fatten themselves at sea before returning ashore for a month to moult, where they lie closely packed together.

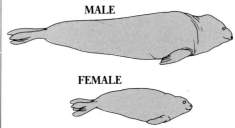

MALE

FEMALE

ELEPHANT SEALS
The difference in size between the male and female southern elephant seals is striking (above). A "beach-master" male may have a harem of up to 30 females. The breeding grounds are on the islands of the sub-Antarctic, particularly South Georgia and the South Shetlands. Elephant seal pups (left) are weaned and hunting for themselves in little more than a month.

Leopard seal (*Hydrurga leptonyx*)
Built for speed, the sleek and slender leopard seal has a large head with an enormous gape and a massive lower jaw, a spotted coat and long, tapered fore-flippers. Females are larger than males, measuring, on average, 3m (10ft) and weighing 350kg (770lb).

They live mainly on the northern edge of the pack-ice but are often found in the sea near penguin rookeries and at some of the sub-Antarctic islands.

Opportunistic predators, they eat a wide variety of prey. Like the crabeater they eat large amounts of krill (37 per cent of their diet), and have tri-cuspid teeth, which interlock to form a strainer. They also eat birds, occasionally penguins, but their more common prey is the young crabeater seals. Many crabeaters bear the scars of a close brush with the leopard seal's large canine teeth.

Solitary animals, little is known of their breeding habits except that they feed from the pack-ice in mid-summer, and the young take part in their annual northward migration.

They have a reputation for cunning and ferocity and there is an account of one attacking a man crossing the ice.

PREDATORY SEAL
The Adélie penguins will not dive while a leopard seal patrols (left). Before eating an Adélie, a leopard seal will thrash it against the water with its powerful jaws (below).

However, it is likely that the seal mistook him for a penguin! Their natural curiosity often leads them to approach small boats and to move towards anyone who comes between them and their escape route to the sea.

Ross seal (*Ommatophoca rossii*)
The least well known and least abundant of the Antarctic seals, Ross seals are also the smallest: females and males measure 2.3m (7ft 6in) long and weigh up to 200kg (440lb).

Their profile is instantly recognizable. They have a very short snout on a wide head and eyes up to 7cm (2½in) in diameter, which bulge under the skin. They also have very long rear flippers which make up more than 20 per cent of their body length. When approached by humans they emit strange clicking and gurgling noises.

Long regarded as a rarity, only 200 had been sighted up to 1970, as they live in the thickest areas of the pack-ice. Believed to be distributed right round the Antarctic region, they are more abundant in certain areas, such as the King Haakon VII Sea and near Cape Adare. They live singly or in small, widely scattered groups. Little is known of their seasonal movements or breeding habits, although their pups are born in November and December. Their main food is believed to be large squid, which they catch with their hooked, needle-sharp teeth.

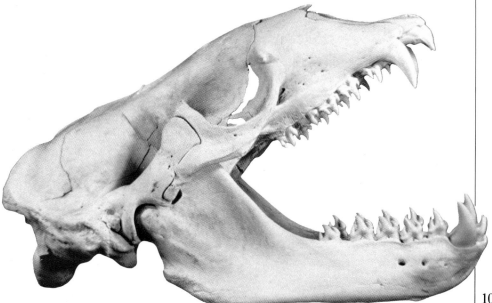

Weddell seal (*Leptonychotes weddelli*)
The most southerly naturally occurring mammal in the world, Weddell seals live on or under (during the dark winter months) the fast-ice near the Antarctic coast all year round.

Large seals with short heads in proportion to their bodies and a short snout, male and female Weddell seals commonly weigh up to 400kg (880lb) and measure 3m (10ft) in length. They eat mainly fish and squid and some crustaceans. Their large eyes are well-adapted to the low light intensities under the ice. They can dive deeply – to 600m (2,000ft) or more – and can stay submerged for more than an hour. Like whales and dolphins, they use sonar to locate their food and to find their way back to holes in the ice.

During the winter darkness they live permanently under the ice, continually gnawing and scraping away at the ice with their teeth to keep open breathing holes. Their teeth become worn down as a result, and their gums often develop abscesses. Very few Weddell seals live longer than about 18 years for this reason. (By comparison, some crab-eaters have lived as long as 39 years.)

During September and November – the Antarctic spring – pregnant females gather on top of the ice around the breathing holes and give birth on the ice. The pups weigh 25kg (55lb) at birth, double their weight in ten days, and start swimming within a month. The mature males defend three-dimensional underwater territories beneath the pupping colonies, and mate with females that enter the water.

OTARIID SEALS
The otariids – the fur seals and the sea-lions – have a visible earflap, and use all their limbs for locomotion on land, which they do very well.

Antarctic fur seal
(*Arctocephalus gazella*)
Compared to all the other Antarctic species, the fur seals are very nimble on land. They can turn their hind limbs forward underneath their bodies, and

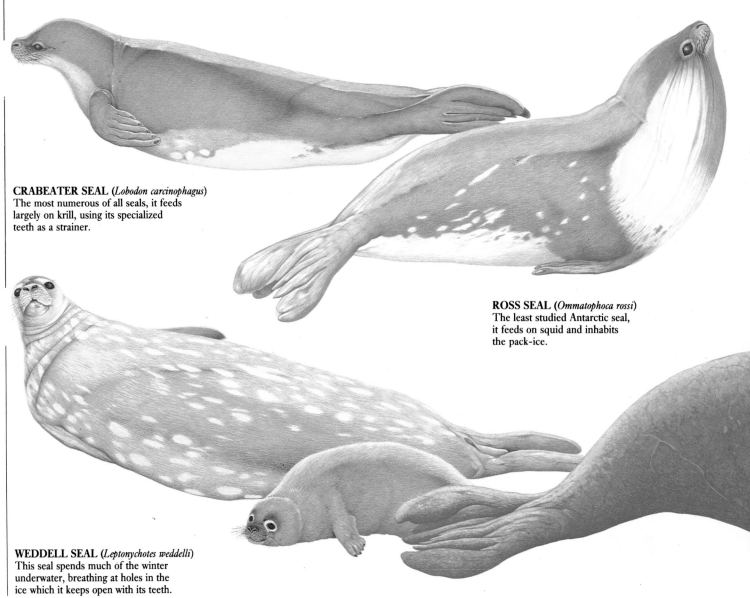

CRABEATER SEAL (*Lobodon carcinophagus*)
The most numerous of all seals, it feeds largely on krill, using its specialized teeth as a strainer.

ROSS SEAL (*Ommatophoca rossi*)
The least studied Antarctic seal, it feeds on squid and inhabits the pack-ice.

WEDDELL SEAL (*Leptonychotes weddelli*)
This seal spends much of the winter underwater, breathing at holes in the ice which it keeps open with its teeth.

balance their weight on their very large fore-flippers. This enables them to walk or even gallop.

Medium-sized males measure 2m (6ft 6in) and weigh 125 to 200kg (275 to 440lb); females are smaller. Apart from being larger, males are recognizable because of their heavy mane around the neck and shoulders.

Their coat consists of two layers: an outer layer of stout guard hairs and an inner layer of very fine fibres which traps air to provide thermal insulation. This inner layer is necessary because they do not have such a well developed layer of blubber as other seal species.

They can regulate their body heat through the black exposed skin of their flippers, either radiating heat from them or, to conserve heat, tucking them under their body.

Found on the sub-Antarctic islands, their main breeding population is on South Georgia. Colonies disperse in the autumn. Their diet consists mainly of krill, which are swallowed whole, other fish and some squid.

Their breeding grounds, to which they return every year, tend to be on sandy beaches. The bulls arrive in late October, several weeks before the females arrive to pup. The cows give birth two days after coming ashore and remain with their pups for eight days, suckling at six-hourly intervals. At the end of this time they mate with the bulls.

The females then go to sea for three to six days to feed and synthesize more milk, and return to suckle for three days more. This pattern is repeated for up to 115 days. The pups, which weigh 5.9kg (13lb) at birth, gain weight at a rate of 98g (3.5oz) a day.

During this period the males spend most of their time defending their territorial boundaries, losing weight at the rate of 23kg (50lb) per day. Their vocal threat consists of a high-pitched whimper. If they fight, the males slash at each other with their sharp canines with which they are capable of inflicting severe wounds. At the end of April they disperse to the open sea until the next breeding season.

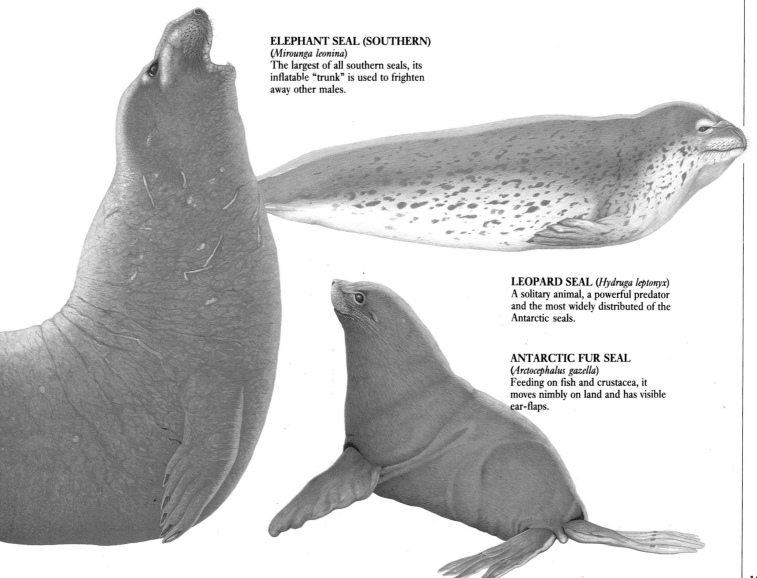

ELEPHANT SEAL (SOUTHERN)
(*Mirounga leonina*)
The largest of all southern seals, its inflatable "trunk" is used to frighten away other males.

LEOPARD SEAL (*Hydruga leptonyx*)
A solitary animal, a powerful predator and the most widely distributed of the Antarctic seals.

ANTARCTIC FUR SEAL
(*Arctocephalus gazella*)
Feeding on fish and crustacea, it moves nimbly on land and has visible ear-flaps.

103

WHALES

Powerful and magnificent creatures, these much hunted mammals are descended from much smaller, four-footed land animals.

The main species of whale found in Antarctic waters are the blue, fin, sei, minke, humpback and right (the baleen or mysticeti whales); and the sperm, orca, and a number of smaller and rarer species, such as the southern bottlenose (the toothed or odontoceti whales).

MYSTICETI WHALES

These whales derive their common name (baleen) from the horny substance, called baleen, contained in plates on either side of the whale's mouth. The plates are fringed with fine hairs (bristles), which form a giant sieve for straining krill and other sea creatures from the water. The size and shape of the bristles varies from species to species; the finer the bristles, the smaller the planktonic animals that can be successfully trapped.

Baleen whales breed in tropical and sub-tropical waters in the north, migrating to Antarctic waters to feed during the summer months on a rich diet of krill. After three or four months of intensive feeding, they migrate back to temperate waters, where the females, a year after mating, give birth to a single calf. The food reserves, accumulated while in Antarctic waters, are stored in their blubber and body tissues to sustain them through the rest of the year.

The calves accompany their mothers on the next migration south, living on their mother's milk. (In the case of the blue whale, calves drink 600l/130gal a day, often doubling their weight in a week on this rich, concentrated diet.) Some six months after their birth, they are able to feed independently in the Antarctic feeding grounds, and to follow the normal migration cycle with the adult whales.

The only exceptions to this rule are the humpbacks and the southern right whales, where calves are dependent on their mothers for up to a year. Smaller minke whales tend to remain in temperate waters all year, although the large adults make the trip to Antarctica.

The various species of baleen whale migrate to the region at different times and some penetrate closer to the Antarctic continent than others. Blue whales arrive first, followed by humpbacks and fin whales, with sei whales the last to arrive.

Baleen whales feed in two ways. In so-called swallow feeding, they engulf a mouthful of food and water and then squeeze the water out by contracting the grooves in their throat and raising their massive tongues firmly against the roofs of their mouths. By contrast, right and sei whales swim through a swarm of krill

PROFILES *The main whale species seen as they "breach" the surface and blow.*

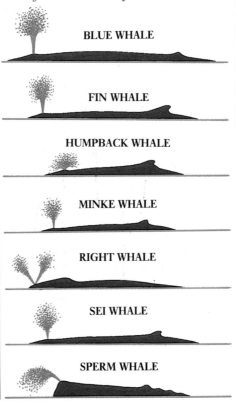

BLUE WHALE

FIN WHALE

HUMPBACK WHALE

MINKE WHALE

RIGHT WHALE

SEI WHALE

SPERM WHALE

with their heads partly out of the water and their mouths half-open, skimming food from the ocean.

Humpbacks have developed an extraordinary feeding strategy. They swim in a circle below the surface, releasing a trail of air bubbles, trapping the krill inside a bubble net. They then swim up through its centre, engulfing the food.

Blue whale *(Balaenoptera musculus)*
The largest of all whale species, the blue whale is probably the largest animal that has ever lived on Earth. It grows up to 30m (100ft) in length and has a maximum recorded weight of 180 tonnes. Blue whales are shy, endangered and rarely seen as they were driven to near-extinction by whaling. There are an estimated 1,000 left of the pre-whaling population of 200,000. It may take a century before there are sufficient numbers to ensure that the species survives.

Fin whale *(Balaenoptera physalus)*
Second only to the blue whale in length and weight, fin whales grow up to 25m (82ft) and to an average weight of 40 to 50 tonnes. Females are slightly smaller. The fin whale is the fastest swimmer of the great whales, and the commonest.

Humpback whale
(Megaptera novaeangliae)
Humpback whales may reach 17.5m (58ft) in length and a maximum weight of 48 tonnes. Their skins are encrusted with barnacles and covered with warts and bumps. It has extraordinarily long flippers, which are up to a third of its total body length. Humpbacks produce the longest and most varied songs in the animal world.

Minke whale
(Balaenoptera acutorostrata)
With an average length of 8m (26ft) and weight of 6 to 8 tonnes, minke whales

BLEACHED SKELETON *The bones of this long-dead blue whale have been reassembled on a beach in Admiralty Bay.*

live for about 50 years and, according to some scientists, seem to be making a remarkable recovery.

Right whale (southern)
(Balaena glacialis or australis)
Fat and stocky, with a smooth, finless back, the right whale weighs a maximum of 96 tonnes. Its large head makes up a quarter of its total length of 18m (60ft) and is decorated with callosities – crusty growths on the skin which contain thriving colonies of barnacles, parasitic worms and whale lice.

Sei whale (Balaenoptera borealis)
Sei whales grow up to 18.5m (61ft) in length and weigh up to 29 tonnes. They have a lifespan of 70 years and it is thought that they may be monogamous.

ODONTOCETI WHALES
These toothed whales differ from other mammals in having only one nostril, although two nasal passages are present.

The sperm whales have a complex social organization. Schools of females and young males live in temperate waters all year and are joined by the males during the summer when mating takes place. Each male has a harem of 20 to 30 females. The calves are born after a gestation period of 15 months and are then suckled for one to two years. The females then rest for nine months, completing the four-year cycle.

Sperm whales are solitary animals while feeding, whereas orcas travel in family pods of five to 20 animals.

Orca whale (Orcinus orca)
Male orcas reach a length of 9m (30ft), while females reach 8m (26ft); the males weigh up to 7 or 8 tonnes. The prominent dorsal fin can measure up to 1.8m (6ft) in height in males, but is less than half that size in females.

Orcas are voracious pack hunters, feeding on squid, fish, seabirds, sharks, seals and other whales. They are particularly fond of the tongues of the large baleen whales and are so strong that they can toss a fully grown adult sea-lion up into the air.

Southern bottlenose whale
(Hyperoodon planifrons)
With a cylindrical body that tapers towards the tail, the southern bottlenose whale has a distinctive "melon" lump on its bulbous head, which contains oil. It reaches 7m (23ft) or more in length and weighs 3 to 4 tonnes. This species of whale is very rarely seen, and it is believed to survive on a diet that consists entirely of squid and cuttlefish.

Sperm whale (Physeter macrocephalus)
Largest of the toothed whales, the male sperm whales can grow up to 18m (60ft) long and weigh up to 70 tonnes, while the females grow to less than 11.5m (40ft) long and weigh up to 17 tonnes. Only the males reach Antarctic waters.

The blunt head of the sperm whale, which makes up one third of its body length, contains a wax-filled "case" on top of its skull. The whale can change the density of this spermaceti wax as it dives, thus altering the whole buoyancy of its body.

Sperm whales feed on more than 40 species of squid, including the giant squid whose sucker marks are to be found scarring the skin of most members of the species. They hunt at great depths, descending at a speed of 7 to 8km per hour (4 to 5mph) to depths of 1,000m (3,300ft) and below, where they can stay for 45 minutes. In these pitch-black waters they locate their luminous prey using sonar.

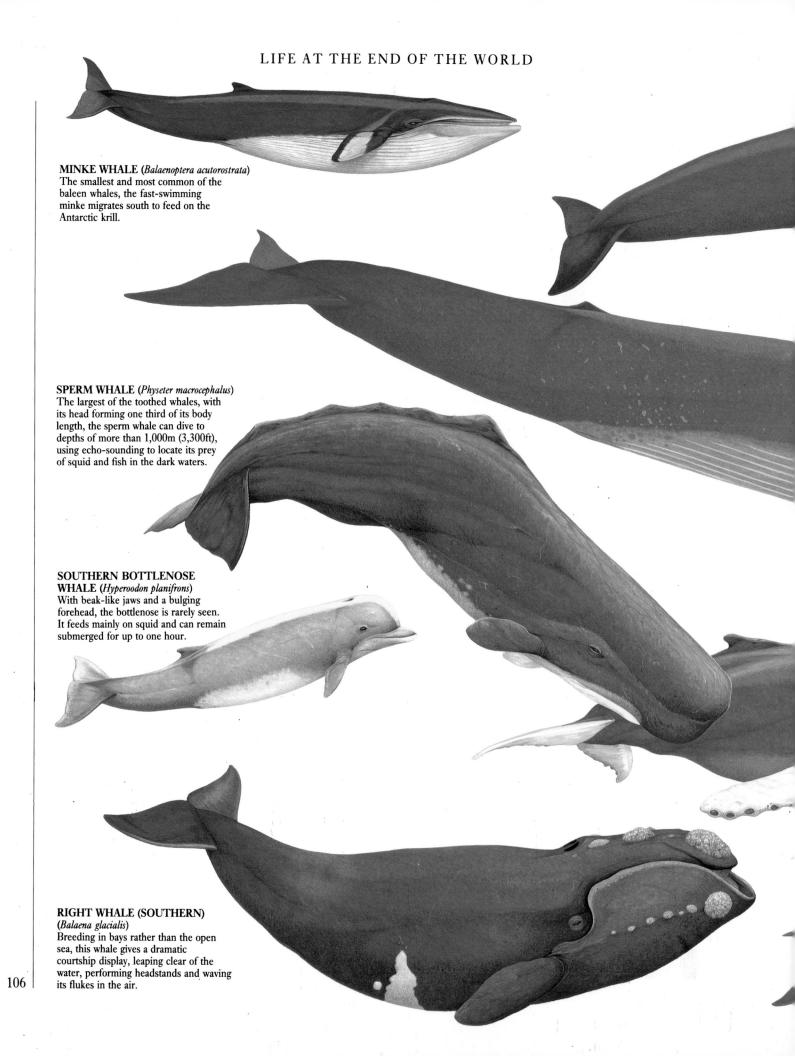

MINKE WHALE (*Balaenoptera acutorostrata*)
The smallest and most common of the
baleen whales, the fast-swimming
minke migrates south to feed on the
Antarctic krill.

SPERM WHALE (*Physeter macrocephalus*)
The largest of the toothed whales, with
its head forming one third of its body
length, the sperm whale can dive to
depths of more than 1,000m (3,300ft),
using echo-sounding to locate its prey
of squid and fish in the dark waters.

**SOUTHERN BOTTLENOSE
WHALE** (*Hyperoodon planifrons*)
With beak-like jaws and a bulging
forehead, the bottlenose is rarely seen.
It feeds mainly on squid and can remain
submerged for up to one hour.

RIGHT WHALE (SOUTHERN)
(*Balaena glacialis*)
Breeding in bays rather than the open
sea, this whale gives a dramatic
courtship display, leaping clear of the
water, performing headstands and waving
its flukes in the air.

FIN WHALE (*Balaenoptera physalus*)
Living and breeding in the open ocean, the fin whale remains north of the ice-edge feeding on krill and small crustacea.

WHALES OF THE SOUTHERN OCEAN

Two orders of whale inhabit Antarctic waters: the baleen and the toothed whales. The blue, fin, sei, minke, humpback, and southern right whale are all from the baleen family. The right whale was so called because it was the "right" one – ie, the easiest to kill. The sperm, orca and bottlenose whales are all toothed whales, which differ from all other mammals in having only one nostril.

BLUE WHALE (*Balaenoptera musculus*)
The largest creature that has ever lived, the blue spends the summer in Antarctic waters feeding on krill. It breeds in mid-ocean.

HUMPBACK WHALE
(*Megaptera novaeangliae*)
So-called because it arches its back sharply before diving, the humpback remains close to coasts on its north–south migrations and breeds in shallow waters.

ORCA WHALE (*Orcinus orca*)
The largest of all carnivores, the orca is found from the tropics to the edge of continental Antarctica, making its way through leads in the pack-ice to feed on seals and penguins.

SEI WHALE (*Balaenoptera borealis*)
The sei rarely enters the pack-ice, staying in deeper waters where it feeds on fast-swimming prey such as Antarctic herring.

THE HUMAN PRESENCE

"I watched the sky a long time, concluding that such beauty was reserved for distant, dangerous places, and that nature has good reason for exacting her own special sacrifices from those determined to witness them."
Admiral Richard E. Byrd – *Alone (1938)*

"We had lived long amid the ice, and we half-unconsciously strove to see resemblances to human faces and living forms in the fantastic contours and massively uncouth shapes of berg and floe."
Sir Ernest Shackleton – *South (1919)*

EXPLORATION

Since 1773, when Captain James Cook crossed the Antarctic Circle, many brave men have been drawn to the most inhospitable region on Earth.

When, in the 15th century, Europeans began to look beyond Europe, they assumed, like the ancient Greeks, that there must be a vast southern landmass – Terra Australis – to balance the northern one. Accordingly they drew up maps that showed a huge southern continent, stretching all the way to the South Pole.

Explorers spent the next century scouring the southern seas for this vast continent. In 1519 Ferdinand Magellan sailed down the coast of South America and into the Pacific Ocean. He saw Tierra del Fuego to the south and assumed that he was looking at the tip of the southern continent.

This theory was not discredited until 1578, when Francis Drake was blown

DISCOVERY
The first British ship to be built specifically for exploration, the Discovery *carried Scott and his expedition to their first assault on the South Pole. Her wooden hull is more than 60cm (2ft) thick, and her bow is strengthened with steel. Seen here with her wind-powered generator buckled by gales, she survived storms, icebergs and imprisonment in the pack-ice.*

KEY DISCOVERIES AND EXPEDITIONS

1773 Captain James Cook and his men cross the Antarctic Circle.

1820 First sighting of the Antarctic Peninsula by Edward Bransfield.

1820 First sighting of the Antarctic continent by Thaddeus von Bellingshausen.

1823 Sealer, James Weddell, travels the furthest south so far – 74°S – entering what is now called the Weddell Sea.

1831 John Biscoe of the sealing company Enderby Bros. sights Enderby Land.

1840 Lieutenant Charles Wilkes sights Wilkes Land.

1840 Discovery of Terre Adélie by Jules Sebastian Dumont d'Urville.

1841 Pack-ice penetrated by Sir James Clark Ross, who discovers Victoria Land, Ross Island, Mount Erebus and the Ross Ice Shelf.

1873 First German Antarctic expedition under the command of sealer, Edward Dallman, discovers the Bismarck Strait.

1895 First landing on the Antarctic continent by whaler, Henryk Bull.

1899 The *Southern Cross* expedition lands at Cape Adare in Victoria Land, and spends a winter on the continent.

1902 The German *Gauss* expedition sights a stretch of Antarctic coast and names it Wilhelm II Land.

1902 Robert Scott, Edward Wilson and Ernest Shackleton make the first attempt to reach the South Pole. At 82°S they are forced to turn back.

1908 Ernest Shackleton, Frank Wild, Eric Marshall and Jameson Adams attempt to reach the South Pole.

They are forced to turn back only 180km (112mls) from the Pole.

14 December 1911 Amundsen reaches the South Pole.

18 January 1912 Robert Scott and his party reach the South Pole.

March 1912 Scott, Wilson and Bowers die 18km (11mls) from the next depot.

1912 Wilhelm Filchner discovers Luitpold Coast.

1914 Ernest Shackleton fails to make the first crossing of the continent.

111

HISTORY PRESERVED

In the constant low temperatures, the stores from Scott's Terra Nova *expedition lie stacked on shelves in his hut on Ross Island, just as they were left more than 75 years ago. Outside the Cape Evans hut, the wind-dried carcass of one of Scott's huskies lies where it fell, the leather collar still intact.*

far south in the Pacific and reported that "there is no main nor island to be seen to the Southwards; but the Atlantic Ocean and the South Sea meet in a large and free scope".

Over the next two centuries sailors continued to explore south of Cape Horn. They discovered Australia, the Falkland Islands, the South Shetland Islands, South Georgia and the Kerguelen Islands.

The myth dispelled

It was Captain James Cook who, after crossing the Antarctic Circle in 1773 and spending a fruitless three years searching the southern seas for the lost continent, finally dispelled the enduring myth of Terra Australis.

Although Captain Cook was certain that there existed no massive southern continent, he declared that he was firmly of the belief that "there is a tract of land near the Pole, which is the source of all the ice spread over this vast Southern Ocean".

Sealing voyages

At the end of the 18th century, scouring the Southern Ocean for land supporting valuable fur seal colonies, the sealers discovered many sub-Antarctic and Antarctic islands.

However, it was the Russian explorer, Thaddeus von Bellingshausen, who made the first sighting of the Antarctic continent in January 1820. In the same month Edward Bransfield of the Royal Navy sighted the Antarctic Peninsula.

Sealers discovered more Antarctic islands and sighted several stretches of Antarctic coastline, including Enderby Land, that were still uncharted.

National expeditions

During 1840 and 1841 three national expeditions took place. The first was a French expedition led by Dumont d'Urville. He discovered a stretch of Antarctic coast, which he named Terre Adélie after his wife.

Overlapping the French expedition was an American one, led by Lieutenant Charles Wilkes. He sighted the part of the coast now known as Wilkes Land.

One year later, in January 1814, a Briton, Sir James Clark Ross, discovered Victoria Land, Ross Island, Mount Erebus and the Ross Ice Shelf, which he likened to the cliffs of Dover. His ships, the *Erebus* and *Terror*, were the first to penetrate the pack-ice.

Ebb and flow of interest

HMS *Challenger*, on a four-year scientific cruise of the world, was the only national expedition to visit the Southern Ocean during the next 55 years.

Then, in July 1895, the International Geographical Congress designated Antarctica as the main region of exploration. Still, the first landing on the continent was claimed by a whaler, Henryk Bull, in 1895. He and his party landed at Cape Adare, and one of the members – Carsten Borchgrevink – returned in 1899 with the *Southern Cross* expedition. They were the first men to over-winter on the continent.

THE WHALING EXPLORERS *This weathered bone from a whale skull remains from the days of the early hunting expeditions, when man first set foot here.*

The years of heroism

The beginning of the 20th century marked the start of man's heroic attempts to conquer the Antarctic continent, to triumph over the awful terrain and severe cold.

In November 1902 Robert Scott, Edward Wilson and Ernest Shackleton made the first attempt to reach the South Pole. They covered 5,000km (3,100mls) and reached 82°S before being forced to turn back.

Ernest Shackleton, Frank Wild, Eric Marshall and Jameson Adams began their attempt to reach the South Pole in 1908. Only 180km (112mls) from the Pole, in poor health and very hungry, they were forced to turn around.

In January 1909 Edgeworth David, Douglas Mawson and Alistair McKay reached the South Magnetic Pole.

Triumph and tragedy

The struggle for the South Pole was finally won by Norwegian, Roald Amundsen, who arrived at the Pole with four others on 14 December 1911.

Just over a month later, on 18 January 1912, Robert Scott, Edward Wilson, Henry Bowers, Edgar Evans and Lawrence Oates reached the South Pole. "Great God!" wrote Scott, devastated at being beaten to the Pole. "This is an awful place and terrible enough for us to have laboured to it without the reward of priority."

They did not complete the journey back. Tragically, at the end of March, Scott, Wilson and Bowers died, only 18km (11mls) from the next depot.

Thwarted expedition

On hearing of Roald Amundsen's success, Ernest Shackleton determined "to start a last great journey".

He returned to Antarctica in October 1915 in the *Endurance* but he never reached the continent. After being locked in the pack-ice of the Weddell Sea for 281 days, the *Endurance* was finally crushed. The party abandoned ship and made for Elephant Island, where they set up camp.

The following April, Shackleton and five of his men set sail from the island in a tiny craft. Sixteen days later, they reached South Georgia and traversed the island on foot, making for the whaling station at Stromness.

Dawn of a new age

Shackleton's expedition failed completely in its object, yet he counted it a success. A large amount of scientific work was carried out, and, he declared, "the comradeship and resource of the members . . . was worthy of the highest traditions of Polar service".

Shackleton died in January 1922 aboard the *Quest* and was buried on South Georgia. The age of exploration died with him and a new age began – the age of political wrangling over territories, and scientific research.

113

BIRTH OF A TREATY

By 1920 the Golden Age of Antarctic exploration was over. There would be other expeditions but none would carry with them the same power and glamour. The unknown had been reduced and the scientist became the key figure in the new era of Antarctic history.

Unfortunately, the explorers left behind a web of interlocking rivalries, of competing territorial claims for national ownership of the great white continent. Great Britain, France, Australia, New Zealand and Norway recognized each other's claims but Chile and Argentina disputed Britain's claims and each other's. In addition, the US reserved the right to make a claim, and there was much suspicion about Soviet plans.

International Geophysical Year

A useful step towards resolving these disputes came in 1950, when a proposal was put forward to celebrate a Third Polar Year (the first was in 1882/83, the second in 1932/33) in order to take advantage of a solar "maximum" – a period when solar activity would be at its most intense.

The idea took off, receiving the sponsorship of the International Council of Scientific Unions. Known as the International Geophysical Year (IGY), its twin aims became to explore outer space (using the new rocket and balloon technology, and Earth satellites) and Antarctica – two unknowns that were now within technological reach.

IGY, actually 18 months long, from 1 July 1957 to 31 December 1958, was supremely successful, involving scientists from 67 countries and a wide variety of disciplines. It became not only the inspiration for numerous other international scientific projects in the 1960s but also led, in due course, to the formation of an international Scientific Committee for Antarctic Research (SCAR), still in operation today.

How SCAR works

SCAR is a non-governmental organization created as a committee of the International Council of Scientific

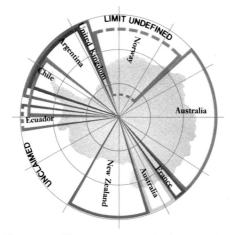

DEFINED TERRITORIES *As they were when the Antarctic Treaty came into force.*

Unions (ICSU). Scientists from about 20 countries participate in its work, largely through 10 permanent Working Groups. Its income is derived from the national organizations in proportion to their Antarctic activity. Its recommendations are not binding.

A new political framework

With the close of IGY in sight, politicians became concerned that nations would use their established scientific bases to advance their territorial claims.

Discussions to provide a new political framework for Antarctica used the 1948 Escudero Declaration proposed by Chile as a starting point. This advocated a five-year moratorium on Antarctic sovereignty disputes to allow scientific work to continue, and endorsed the principles of free access to the continent and political neutrality for expeditions.

With this in mind, representatives of 12 nations met bi-weekly from June 1958 until early in 1959 at the National Science Foundation in Washington. The resultant Antarctic Treaty was signed on 1 December 1959 and came into effect on 23 June 1961.

Main points of the Treaty

The linchpin of the Treaty is Article IV which, in effect, recognizes that the question of territorial sovereignty is insoluble. Its ambiguous meaning has been variously interpreted in law as a first step either towards *terra communis*, coownership of Antarctica, or *terra nullius*, renunciation of ownership.

Article IX permits other countries to become Consultative Parties if they are engaged in "substantial scientific research activity" and 20 nations are now represented.

In addition, a second level of non-voting participants – acceding states, or Non-Consultative Parties – has now developed. These are countries which, while not engaged in substantial research, agree to abide by the terms of the Treaty and ratify it.

Consultative Meetings

The formal framework of Treaty cooperation is provided by the Antarctic Treaty Consultative Meetings (ATCMs), which are generally held every two years. Decision making is often slow at these forums, so negotiations on important issues are often carried out at Special Consultative Meetings. Until recently, ATCM documents were kept secret.

An international continent

The Antarctic Treaty has had the practical effect of stabilizing the conflicting political aims of a truly international continent. Martin Baker, former logistics officer for the British Antarctic Survey, has compiled the following short history of each nation's involvement in Antarctica, together with a breakdown of their current scientific programmes. More detailed information concerning the bases can be found in the Appendices.

ARGENTINA

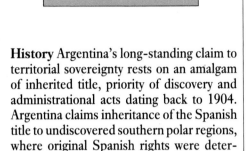

History Argentina's long-standing claim to territorial sovereignty rests on an amalgam of inherited title, priority of discovery and administrative acts dating back to 1904. Argentina claims inheritance of the Spanish title to undiscovered southern polar regions, where original Spanish rights were determined by the Spanish/Portuguese Treaty of Tordesillas, which was drawn up in 1494.

Argentina claims to have evidence of the voyages of the Argentine Admiral, Guillermo Brown, who in 1815 claimed to have seen ice-covered islands at latitude 65°S, at least four years before the voyages of Thaddeus von Bellinghausen (USSR), Nathaniel Palmer (US) and William Smith (UK) to the same region.

On 7 July 1904, the British Government handed over the winter quarters of William Bruce's Scottish National Antarctic Expedition (1902-4) to Argentina for use as a meteorological station, and Argentina renamed the base Orcadas. It is the oldest of all surviving Antarctic stations.

When, in 1927, the UK asked Argentina to apply for a British licence to operate a meteorology station and wireless/post office on "British territory", Argentina replied that Argentine jurisdiction, and not British, applied to Orcadas, and that a licence was therefore not necessary.

Argentina defined the current limits of her Antarctic claim in 1946, as covering the region from 25°W to 74°W. This action came in response to a rapidly growing number of British and Chilean bases in the Antarctic Peninsula region. Science was, prior to the IGY, of secondary importance. Early Argentine military bases on Gamma Island (1947) and Deception Island (1948) maintained simple meteorological and biological programmes only.

Current activities Almost all Argentinian bases are under the explicit control of the Armed Forces and are staffed principally by services personnel. Marambio (on Seymour Island), Argentina's largest base, has about 30 buildings, and a large compacted runway from which the Argentine Air Force operates Lockheed Hercules C130 transport aircraft. Marambio's population includes women and children.

For the month of August 1973, Marambio was declared by Government Decree-in-Law to be a provisional seat of government, and the entire Argentine Cabinet conducted their daily business there. Marambio is more like a frontier settlement than a scientific base, and is increasingly important as an Antarctic tourist centre. There are plans to install hotels, banks and shops, and extensive tourist facilities. In 1978, Emilio Marcos Palma became the first baby to be born in Antarctica, a part of Argentina's declared process of major "Antarctic colonization".

Argentina's Antarctic programmes include most scientific disciplines, although published scientific output is low for the very high level of support activity. Sea-borne research has concentrated on krill and fin fish stock assessments in the North Weddell Sea, and on marine geophysics research aimed at increasing knowledge of the sedimentary basin structures.

ANTARCTIC SHRINE *A statue of the Virgin Mary graces Paradise Bay.*

AUSTRALIA

History The early history of Australia is closely linked to the opening up of the southern polar regions; both were the focus of the intense European search for an unknown southern continent in the 18th and 19th centuries. This search received further impetus from the great sealing and whaling operations of the 19th century.

Special Australian interests in Antarctica proper were formulated in the 1880s after the Challenger Expeditions (1872-76). The earliest Australian scientific stations in Antarctica were temporarily opened in Commonwealth Bay, and on Macquarie Island in the sub-Antarctic, in late 1911 as part of Douglas Mawson's Australian Antarctic Expedition (1911-13). He later organized and led the British-Australian-New Zealand Antarctic Expedition in the summers of 1929 and 1931.

In 1933, Australia inherited from the UK the sovereign claim over what is now known as Australian Antarctic Territory, south of 60°S and between the latitudes of 45° and 160°E, but excluding the French claim of Terre Adélie. The Australian claim therefore incorporates about 40 per cent of Antarctica. In 1947, the Australian National Antarctic Research Expedition (ANARE) was established to coordinate Antarctic programmes, and in 1954 ANARE constructed Australia's first permanent Antarctic mainland station. The station was named after Sir Douglas Mawson.

SLEDGE-DOG *The role of the husky has been crucial in the exploration of Antarctica.*

Current activities Australia's Antarctic priorities are to maintain its sovereign claim over Australian Antarctic Territory, to strengthen the Antarctic Treaty and the consultative framework, and to maintain a balanced scientific programme as a contribution to world science and in support of the Australian claim and the Antarctic Treaty System.

In 1980 the Australian Government announced a major 10-year rebuilding programme for Mawson, Casey and Davis stations. Its 1985/86 overall budget of A$42.0 million was the largest Australian Antarctic effort since the mid-1960s.

Logistics support is provided by the 2,206 tonne supply ship, *Nella Dan*, and the 4,000 tonne ice-strengthened, multi-purpose supply vessel, *Ice Bird*, both chartered on an annual basis. *Nella Dan* concentrates on marine science studies, including krill and fin fish studies, hydroacoustics and general oceanography, and was involved in a seismic survey cruise in Prydz Bay in 1981/82. There are plans for a new ice-strengthened research ship for marine geosciences to explore the Antarctic continental shelf.

Proposals for air traffic facilities capable of taking Hercules C130 transport and Boeing 747 planes could place Australia at the forefront of air traffic into Antarctica. Australia has recently increased her commitment to research the region's economic potential, including hydrocarbon exploration, krill harvesting and iceberg utilization.

BELGIUM

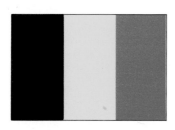

History Belgium quotes a long-standing interest in Antarctica, dating back to 1897 and the exploits of Royal Belgian Navy lieutenant, Adrian Victor Joseph de Gerlache. De Gerlache led an expedition to the west coast of the Antarctic Peninsula in 1897-99, during which his ship, the *Belgica*, became trapped in ice for 13 months.

Belgium did not return to Antarctica until the IGY. Baron Gaston de Gerlache de Gomery, the son of Adrian de Gerlache, helped to organize the Belgian contribution, which included the establishment of King Baudouin base. Research was conducted from this station from 1957 to 1961. Belgium and the Netherlands jointly organized a new expedition in 1963.

In 1964, 14 winterers (including four Dutch scientists) pursued studies in geology, geodesy, oceanography, biology, atmospheric sciences, glaciology, geomagnetics and, of course, meteorology. Joint Belgian/Dutch activities continued until February 1967, when the base closed again. In its last summer, one Italian and two Spanish scientists also participated. From 1967 to 1970, Belgium concentrated on cooperation with the research programme being run by South Africa.

Current activities Since 1970, the Belgian contribution has been limited to the exchange of scientists with the US, France, West Germany, Japan and Australia. Belgium currently contributes to two such collaborative programmes, both of which contain a strong element of marine biology and oceanography.

Belgium's position as a Consultative Party is rather anomalous. As an initial signatory to the Treaty, it has permanent Consultative Status, but it does not maintain an Antarctic base, nor an active, independent scientific programme.

BRAZIL

History Although Brazil formally supported India's 1956 initiative to place Antarctica under UN trusteeship, it showed no further interest in Antarctica until 1975, when it acceded to the Antarctic Treaty. Coincidentally, a number of Brazilian geopolitical writers began advocating a "frontage theory" for sovereign claims to Antarctica, by which only those countries with south-facing coastlines projecting on to Antarctica could claim Antarctic territory. If accepted, the frontage theory would grant Brazil a sizeable "South American" sector of Antarctica, at the expense of traditional rivals Chile and Argentina. In effect, Brazil's interest in Antarctica is at least partly a corollary to her desire to project political and economic influence into the strategically important South Atlantic.

In 1982/83, Brazil's first Antarctic expedition utilized two ships, the 2,200-tonne Brazilian Navy ship *Barao de Teffe* (formerly the Danish-built *Thala Dan*), and the 700-tonne University of Sao Paulo research vessel, *Professor W. Besnard*, in support of BIOMASS. After a second expedition a year later, Brazil was granted full Consultative Status to the Antarctic Treaty on 12 September 1983, at the same time as India.

On 2 February 1984, Brazil established its first summer-only base, Commandante Ferraz, on Punta Plaza, in Admiralty Bay, on the already crowded King George Island in the South Shetlands.

Current activities In 1984/85, Commandante Ferraz was expanded to accommodate over 20 scientists and support staff. Ferraz is now operational all-year round, with programmes of meteorology, upper atmospheric and cosmic ray physics, and geomagnetics. Twelve men and women over-wintered in 1986. Two summer-only refuges (Astronomo Cruls and Engenheiro Wiltgen) also support geological and biological research studies.

CHILE

History Chile's Antarctic interests closely parallel those of its rival, Argentina. Chile's claim is also based on a mixture of historical inheritance, geographical contiguity, ecological interdependence and supportive acts of regional administration. One of the earliest manifestations of a "Chilean Antarctica" concerns Charles V of Spain, who in 1539 allegedly granted the adventurer Pedros Sanchez de la Hoz the territory "from the Strait of Magellan to the South Pole inclusive", west of 40° longitude. Chile argues that de la Hoz and his successors (eventually to include the Captaincy General of Chile) were bona fide governors of Antarctica, albeit 280 years before the continent was unequivocally discovered. In 1831, the first President of the Republic, General Bernardo O'Higgins, in a farsighted act of international diplomacy, submitted a letter to the British Government confirming that Chilean territory extended at least to the 65°S latitude, and perhaps by implication to the Pole itself.

Chile has also produced evidence in support of priority of discovery, by a Chilean sailing ship in 1820, to strengthen her claims to sovereignty.

Chile established a whaling station on Deception Island in 1906, but it was not until 1940 that it formally laid claim to Antarctic territory (53° to 90°W). This effectively tripled the size of Chile, and overlapped not only the British claim, but also one that Argentina subsequently made. In 1947, Chile established a permanent base – Soberania – on Greenwich Island as part of a large Chilean naval presence in the region. A year later, President Gonzalez Videla became the first head of any state to visit Antarctica; he inaugurated a second Chilean base – Bernardo O'Higgins. A third base was built in February 1951.

Since the signing of the Antarctic Treaty, which it played a large part in drafting, Chile has supported the Treaty whilst striving to extend her presence and political authority in and around the Antarctic Peninsula. In 1969 Presidente Frei Montalva was opened as a regional meteorological centre on King George Island.

In 1977 and 1984 Chilean President Pinochet visited Chilean Antarctic Territory. In 1980 Chile constructed an airstrip on King George Island, and the associated settlement – Teniente Rudolpho Marsh – has recently been supplemented with families, shops, a bank and tourist facilities, in the spirit of the Argentine base Marambio. The settlement is known as – "Villa Las Estrellas" – Town of the Stars.

Current activities Chile presently operates five permanent stations, of which three are open all year. In addition, several refuge huts are available to support occasional summer programmes. Support for science and logistics includes four ships, several helicopters and some fixed-wing aircraft.

The relief of Teniente Rudolpho Marsh, with its airstrip, takes place in November by Chilean Air Force C130s. Thrice-weekly two-hour flights to and from Punta Arenas in southern Chile make Marsh base one of the most frequently connected to the outside world, and therefore one of the most important ones.

By 1986, nine major stations representing seven nations were located on King George Island, itself only $1,450km^2$ ($900mls^2$) in area. Marsh base is an increasingly popular tourist site – a role actively encouraged by the Chilean Government. In October 1982 it hosted an international conference on Antarctic politics, the first conference to be held on the Antarctic continent itself.

Geological exploration in the Antarctic Peninsula region is conducted under the auspices of Chile's national oil company.

CHINA

History China acceded to the Antarctic Treaty on 8 June 1983, thereby signalling an intention to become actively involved in the management apparatus of Antarctica. It was a significant development. China's recognition of the Treaty principles was welcomed by the other Treaty states, which hoped that it would set an example to the Third World nations, who view the Treaty with distrust.

China's first Antarctic expedition in 1984/85 established a year-round station – Chang Chen (Great Wall) – on King George Island, about 1km (0.6mls) from the Russian base, Bellingshausen, and the Chilean base, Presidente Frei. An astonishing 591 crew, construction workers and scientists constituted one of the largest summer operations of any nation, and certainly the largest "maiden programme" in Antarctic history. Several neighbouring bases reported a lack of awareness of the principles of the Treaty, especially the Agreed Measures, among the construction workers. Other states' scientific programmes were also disturbed.

Unofficial Chinese interest in Antarctic research began in the late 1970s, when several Chinese scientists participated in the research programmes that were being conducted by Australia, New Zealand, Argentina, Chile and Japan.

Current activities China returned to Chang Chen in November 1985, having been granted Consultative Status in the Antarctic Treaty on 7 October 1985. Thirty scientists conducted research programmes. China's main research interest is marine science, and the *Xiang Yang Hong* research ship provides a platform for the study of marine living resources, particularly krill, as part of the Chinese contribution to the Biological Investigation of Marine Antarctic Systems and Stocks.

FRANCE

History French interests in Antarctica date back to the 16th century when the French navigator, Palmuyer de Gonneville, returned from two years of exploration with extraordinary tales of a previously undiscovered southern tropical paradise. The legend of "Gonnevilleland" persisted for two centuries, and in 1739 Jean Baptiste Charles Bouvet de Lozier, sent to establish new trading posts in the Far East and to explore new southern regions for Gonnevilleland, discovered the bleak uninhabited sub-Antarctic island that is now known as Bouvet Island.

Further expeditions – Kerguelen (1769), Dumont d'Urville (1840) and Charcot (1903 and 1908-10) – served to demonstrate France's sporadic interest in Antarctica. In 1911, the UK wrote to France requesting clarification of France's role in Antarctica, presumably so that the UK could avoid any possible embarrassment that might arise in the pursuit of her own considerable territorial ambitions.

In April 1912, France replied that Dumont d'Urville had taken formal possession of "Terre Adélie" in 1840, although no specific limits to the claim were mentioned. These were formally identified by French decree on 27 March 1924.

In 1947 the French Ministry of Education established the "Expéditions Polaires Françaises" to organize the first modern French expedition to the Antarctic. Seventeen men wintered in Terre Adélie in 1951, on a base constructed at Port Martin. A second base was constructed, this time on Pointe Géologie on Petrel Island (the site

ASTROPHYSICS STUDIES *Atmospheric conditions make this an ideal research site.*

for Dumont d'Urville's proclamation of sovereignty 111 years earlier), after a fire destroyed Port Martin. The modern base of Dumont d'Urville was erected on Pointe Géologie and opened on 1 April 1956 as part of the IGY, and is the flagship of French Antarctic activities.

Current activities In 1986, 28 personnel wintered at Dumont d'Urville, with a summer complement of over 60. A joint French/American ocean/atmosphere/ice interaction study in the coastal vicinity of Pointe Géologie involved about 20 French and five American scientists, and several automatic weather stations. France also maintains sizeable research activities on the sub-Antarctic islands of Kerguelen, Crozet and Amsterdam.

French plans to build an 1,100m (1,200yd) all-weather airstrip by linking five small islands (Cuvier, Lion, Pollux, Zeus and Buffon) near Pointe Géologie were strongly criticized in 1982 when preliminary construction work threatened local Adélie and emperor penguin colonies, as well as breeding sites for eight different types of flying seabird. Small-scale earthworks continued until 1984 and were then suspended to allow environmental impact assessments.

In early 1986 an Australian inspection team confirmed that the French authorities had postponed the completion of the project. It is likely, however, that construction will begin again in the 1987/88 season.

GERMANY

History Germany conducted early research in Antarctica with an over-wintering party on Kerguelen Island in 1874, and with participation in the First International Polar Year in 1882/83, during which the German South Polar expedition wintered on South Georgia. Station-based investigations were renewed in 1901-4, when an expedition led by Drygalski renewed the scientific station on Kerguelen Island. The expedition ship *Gauss* wintered in the Davis Sea, trapped for several months within sight of the coastline of Wilhelm II Land. In 1911/12, Wilhelm Filchner led a further German expedition.

In May of 1938, Reichsfieldmarschall Hermann Goering agreed to support a new German expedition to Antarctica, which involved the novel use of ship-based aircraft. The aim of the expedition was to learn as much as possible about the South Atlantic regions as quickly as possible, in support of Nazi Germany's whaling ambitions in the area.

Germany also had territorial aspirations, and the "Schwabenland Expedition" laid claim to as much of "Atlantic" Antarctica as possible. Hitler was delighted with the success of the expedition, but the German sovereign claim was subsequently lost in the chaos of World War II.

WEST GERMANY (FRG)

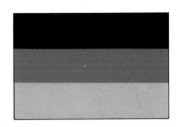

West Germany joined SCAR in May 1978, and acceded to the Antarctic Treaty on 5 February 1979. She opened SCAR's first permanent Antarctic research station – Georg von Neumeyer – on the Ekstrom Ice Shelf, on 24 February 1981. The Filchner summer station was opened in February 1982. Drescher, West Germany's second summer-only station in the Weddell Sea, was deployed in February 1987. West Germany achieved Consultative Status within the Treaty system on 23 September 1983.

Current activities West Germany currently operates three fully operational research stations. Great emphasis is placed on resources-orientated research, with considerable effort devoted to marine and terrestrial geophysics and marine biology. Several West German scientists work regularly on King George Island.

EAST GERMANY (GDR)

East Germany's Antarctic policy is influenced by that of the USSR. Following Russian advice, it acceded to the Treaty on 19 November 1974, and joined SCAR on 9 September 1981, thus bolstering Eastern European representation. East Germany has officially supported a scientific presence in Antarctica since 1976 and its base, Schirmacher Oasis, is now fully independent of the nearby Russian base, Novolazarevskaya, barely 1km (0.6mls) away. East Germany was granted Consultative Status at the 1987 Treaty meeting in Rio de Janeiro.

Current activities East German scientists regularly work at the Russian bases of Bellingshausen, Vostok, Mirny and Molodezhnaya. Schirmacher Oasis supports programmes of meteorology, atmospheric chemistry, atmospheric physics, geophysics, geology and glaciology.

INDIA

History Although India exhibited an awareness of Antarctica as a political issue as early as 1956, when it initiated an unsuccessful move to have Antarctica placed under UN trusteeship, it undertook no activity until 1981. India's reluctance perhaps stemmed from its self-perceived role as a leader of the Third World, which has traditionally viewed the exclusive nature of the Antarctic Treaty System with some degree of suspicion.

The first Indian expedition, of 21 scientists and technicians, set sail from Goa on 6 December 1981, but India did not accede to the Treaty until 19 August 1983, on the understanding that it would receive full Consultative Status (which it did a mere 23 days later). India's decision to join the Treaty was not received enthusiastically by Non-Aligned Movement members.

India installed a wooden surface station on the Prince Astrid coast of Dronning Maud Land, near the Schirmacher Hills, in January 1984. Twelve men over-wintered in 1984, and the new base, named Dakshin Gangotri, became fully operational during January 1985.

Current activities About 15 scientists and support staff currently winter at Dakshin Gangotri, monitoring programmes of meteorology, biology and geology. Only three scientists wintered in 1986. Geologists working in the Schirmacher Hills have located trace occurrences of uranium, copper, lead and zinc. Marine geophysical surveys have been undertaken to determine the mineral potential of the continental shelf margins of Dronning Maud Land.

India is planning an expansion of Dakshin Gangotri, the installation of a summer base on the coast, and the construction of a new summer-only base in the Schirmacher Hills. It also plans an aerial survey of the Wohlthat Mountains, the installation of wind energy and solar power systems at its bases, experimental krill fishing and processing, and a third permanent station.

ITALY

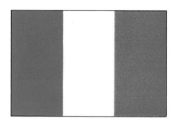

History Italy became the 24th country to accede to the Antarctic Treaty, on 18 March 1981. In 1985, Italy announced that it planned to spend in the region of $US 130 million on Antarctic scientific research in the following five years, a staged commitment greater than that proposed by the UK over the same period. Italy's first official independent expedition took place in the austral summer of 1985/86, when a team of 23 scientists and technicians went to Terra Nova Bay in the Ross Sea. Scientific activities included geophysics, geology, petrology, geochemistry, glaciology, meteorology, atmospheric research and biology. Sites for a forthcoming permanent station were reconnoitred.

Italy's Antarctic history had, up to 1985, been intermittent and under-supported. Italian scientists accompanied Borchgrevink (1898) and Charcot (1903). During the IGY, an Italian naval officer spent eight months at the New Zealand Scott base, thereby beginning a relationship between Italy and New Zealand that was instrumental in Italy's decision to build its first research station in the Ross Sea area in December 1986.

Three Italian expeditions worked in the Ross Sea area with New Zealand logistic support (1968/69, 1973/74, 1976/77).

There were also several private Italian expeditions to the Antarctic Peninsula (1970/71, 1975/76 and 1977/78).

In 1986/87, the Italian National Research Council (CNR) agreed to the establishment of a permanent Italian station in Terra Nova Bay. Thirty-nine scientists and support staff coordinated an ambitious programme of cosmophysics, meteorology, atmospheric physics, geology, geomorphology, geomagnetism, oceanography, biology and pollution studies. The expedition terminated at the end of March 1987.

Current activities As a result of this determined endeavour, Italy has been granted Consultative Status within the Antarctic Treaty System, and it is expected that Terra Nova Bay will eventually become a winter station. This means that, of the seven major western economic powers, six are Consultative Members.

JAPAN

History In an era when it has no tradition of global exploration, Japan's first expedition to Antarctica sailed from Tokyo on 1 December 1910, under the command of Lieutenant Nobu Shirase. His ultimate intention was to achieve the Pole itself, in competition with Scott and Amundsen. Shirase arrived in Antarctica in December 1911, too late to pose a serious challenge to the two main contenders, but he managed to reach 80°05′S in a quick dash, and then withdrew. He returned to a tumultuous welcome in Tokyo, very different indeed from the total lack of interest exhibited at his departure. Japan now perceived itself as having a role in the future of Antarctica.

On 13 November 1940, Japan indicated that it considered itself one of those countries which had unspecified rights in Antarctica, a declaration that should be judged in the context of Japan's then imperialist ambitions and its rivalry with the US, itself poised on the verge of an Antarctic territorial claim.

On 8 September 1951, at the Treaty of Peace signed in San Francisco, Japan was forced by the US to declare that "Japan renounces all claims to any right or title to or interest in connection with any part of the Antarctic area, whether deriving from the activities of Japanese nationals, or otherwise". Japan is thus the only nation constrained by Treaty never to register an Antarctic claim.

Japan had considerable interest in Antarctic regions as a consequence of her whaling industry. Japan's contribution to the IGY consisted of a scientific station on East Ongul Island, off the coast of Dronning Maud Land, part of Norway's sector. Syowa station was opened on 29 January 1957, and, apart from the winter of 1958, has been open ever since. Japan is an original signatory of the Antarctic Treaty. A second base (Mizuho) was opened in July 1970, and a third (Asuka) in January 1985. Asuka was initially designed to be a summer-only geology camp, and was occupied for two months by eight scientists.

Japan ceased operating two land-based whaling stations on South Georgia in the middle to late 1960s. Intensive pelagic whaling ceased in 1987, in respect of the International Whaling Commission indefinite moratorium on commercial whaling. However, Japan signalled its intention to continue with scientific whaling. Japan was also responsible for the first commercial krill-harvesting operation, off Enderby Land, in 1973/74.

Current activities Japanese winter studies tend to concentrate on atmospheric studies and glaciology, while summer research efforts are resources-orientated. Considerable effort is given to krill-based ecological studies, both on the Shirase and at Syowa. Japan has also been responsible for several offshore seismic surveys, using the Japanese Metal Mining Agency's geological survey ship *Hakurei Maru*, sponsored by the Japanese National Oil Corporation (JNOC) – in the Bellingshausen Sea (1980/81), in the Weddell Sea, Ross Sea and off Wilkes Land (1982/83), Prydz Bay (1984/85) and off Queen Maud Land (1985/86). All are sites of sedimentary basins of possible hydrocarbon potential. Further geoseismic cruises are planned.

Asuka became Japan's third all-year Antarctic station in the winter of 1987, with the over-wintering of eight geoscientists and technicians.

NEW ZEALAND

History The New Zealand claim, the "Ross Dependency", dates from a British Order-in-Council of 30 July 1923, making it the second-oldest formally registered claim. However, apart from involvement in the British-Australian-New Zealand Antarctic Research Expedition (BANZARE) of 1929-31, New Zealand's early activities were nominal, and the validity of her claim looked to be in question. Yet New Zealand had a strategic interest in ensuring that Antarctic territories immediately to the south did not fall into unfriendly hands.

New Zealand established a scientific base on Ross Island in the Ross Sea in January 1957, as a timely contribution to the IGY. The base, called Scott base, subsequently served as the Pacific anchor for the 1956-58 Trans-Antarctic expedition, under the joint leadership of Englishman Vivian Fuchs and the New Zealander Edmund Hillary.

American interest in the highly desirable Ross Sea area has always been high, and New Zealand's relations with America over the region have been very close since the IGY. Although the US does not recognize New Zealand's claim, the two nations'

HUMAN STUDIES *University scientists carry out psychological tests.*

activities are logistically interdependent and the US operates a coordinating office in Christchurch. The American McMurdo station, the largest of all Antarctic installations, is within sight of the New Zealand, Scott base.

Current activities Every summer, up to 300 personnel are involved in the implementation of the New Zealand Antarctic Research Programme (NZARP).

Recent new programmes include glacial studies in the Marshall Valley, off-shore core-sampling and sea-floor photography in McMurdo Sound, and work on influenza viruses in seals and birds. A major study of the geological evolution of North Victoria Land has been conducted since 1958, and is presently part of a joint American-German-New Zealand effort.

New Zealand geologists are also studying the coal reserves of the Beacon Supergroup formation. As well as Scott base, New Zealand also operates two summer stations – Vanda and Cape Bird. A major rebuild of Scott base, begun in 1976, should be completed by 1990 and will serve New Zealand interests well into the next century.

DOG DAYS *The sun rises on the New Zealanders' dog team.*

NORWAY

History Norwegian whalers began operating in the Southern Ocean in the middle of the 19th century. In about 1870 the invention of the explosive harpoon revolutionized the whaling industry. Norwegians began whaling from land stations in and around Antarctica in 1906, introducing factory ships and pelagic whaling in 1926 in an effort to avoid British licensing of land stations in the South Atlantic. By the mid-1930s, Norway and Britain shared over 90 per cent of the region's whale catch. By the 1950s, the total annual whale catch from the Southern Ocean was about 30,000 but by the mid-1970s it was down to less than 2,000 per year, and Norway ceased whaling in Antarctica in 1968.

During the 1890s, the work of Norwegian whalers, especially Captain Carl Larsen, was instrumental in the renewal of scientific interest in Antarctica. Another Norwegian, Carsten Borchgrevink, a member of a Norwegian whaling expedition into the Ross Sea area in the ship *Antarctic*, is considered to be the first man to set foot on the Antarctic mainland, on 24 January 1895.

Norwegians, including a young 25-year-old Roald Amundsen, were influential members of de Gerlache's ill-fated *Belgica* expedition to the Antarctic Peninsula (1897-99). Twelve years later, Amundsen returned, not only to become the first man to reach the South Pole, but also to establish Norway's first wintering station, Framheim, in the Ross Sea area.

Norway's interests in what was to become her Antarctic claim developed during the late 1920s and 1930s, again as an offshoot of her whaling activities. Expeditions in 1929/30, 1930/31 and 1936/37 mapped and photographed much of the coastline immediately to the east of the Weddell Sea, as far as Prince Haralds Land.

When Nazi Germany announced the territorial motives of the Schwabenland expedition, Norway quickly preempted a German claim to sovereignty by making one

HAZARDOUS CONDITIONS *The Hägglund finds its way in a "white-out".*

herself, for lands between 20°W and 45°E. Norway had already claimed Bouveya (1928) and Peter I Island (1931). The claim was made in an attempt to protect Norwegian whaling interests along the coastline of Dronning Maud Land.

In 1950, a joint Norwegian-British-Swedish expedition established the winter station of Maudheim, which closed in 1952. The base opened in 1956 as part of her IGY effort was handed over to South Africa in February 1962. This left Norway with considerable involvement in Antarctica but no permanent base there. Nonetheless, summer expeditions in 1970/71, 1976/77, 1978/79 and 1984/85 have contributed significantly to the growing wealth of Antarctic scientific research.

Current activities Norway's summer expedition in 1984/85 was her largest for 25 years. Seventy-seven men and women, including 28 scientists and engineers, took part in programmes of marine science, geophysics, geology, glaciology, iceberg studies, terrestial ecology/invertebrate biology, petrographic studies and ornithology. A geologist from the Norwegian state oil company accompanied the expedition. The *Andenes*, a new Norwegian ice-strengthened vessel, conducted a seismic survey of the southern Weddell Sea margin. Norwegian mainland science was conducted from two camps: Norway 5 (72°S, 5°E) and Norway 6 (75°30'S 10°W). Norway intends to return to Dronning Maud Land in 1987/88, and may install a permanent station there.

POLAND

History Poland's first Antarctic expedition to the Bunger Hills on the coast of Wilkes Land in 1958/59 was organized and financially supported by the Polish Academy of Sciences. The logistics, however, were supplied by the USSR (although paid for by Poland). Thereafter, Poland tried to demonstrate that her Antarctic commitment was worthy of independent consideration, not having been invited to the 1959 Washington Conference responsible for drafting the Antarctic Treaty. In its first expedition, Poland used the Russian IGY base, Oaziz, which was transferred to Poland on 21 January 1959, and renamed Dobrowolski. Early science programmes concentrated on geophysics, atmospheric CO_2 measurements, geomorphology and glaciology. Dobrowolski closed in 1961.

Although denied a part in drafting the Antarctic Treaty, Poland quickly acceded, on 8 June 1961. Several teams of Polish scientists accompanied Soviet Antarctic Expeditions between 1964 and 1970, and in 1971/72, Polish biologists worked on board the Russian research ship *Academik Knipovich*. A Polish biologist spent the 1973/74 summer season at McMurdo base.

By 1975, Polish trawlers had begun to fish extensively in the South Atlantic, prompting the Polish Sea Fisheries Institute to despatch their research ship *Professor Siedlecki* on several cruises to the Scotia Arc and Antarctic Peninsula between 1974 and 1980. Polish krill fishing began in 1976-77. The *Professor Siedlecki* devoted much effort in investigating krill harvesting and krill processing techniques.

ARCTOWSKI STATION
Whale ribs flank the flags on King George V Island.

Poland established its first permanent station on 26 February 1977. Henryk Arctowski base (named after the Polish scientist who accompanied de Gerlache's *Belgica* expedition in 1897), was built in Admiralty Bay on King George Island. As a result, on 29 July 1977, Poland became the first nation to achieve Consultative Status within the Antarctic Treaty System since its original ratification in 1961.

Current activities In 1987, Arctowski housed 19 winterers, including nine scientists, on programmes of meteorology, glaciology, biology, human biology , geology, seismology and petrology. Poland still uses the *Professor Siedlecki* for marine science and logistics support. The *Jantar* is also used for marine geological work, mainly in the Bransfield Strait, South Sandwich Islands, and Antarctic Peninsula. The Soviet Union reopened Dobrowolski in 1986/87, giving it back its old name, Oaziz.

SOUTH AFRICA

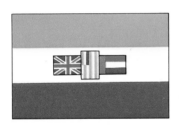

History South Africa's first involvement in Antarctica surprisingly (considering her geographical proximity) did not occur until 29 December 1947. A secret military expedition raised the Union flag over Marion Island, a sub-Antarctic island nominally in the possession of the UK, and six days later over Prince Edward Island. Press reports at the time speculated that South Africa's real intention was a sovereign claim to the Antarctic mainland due south of Africa (an area already claimed by Norway), and that South Africa wished to curtail the growing strategic activities of the Soviet *Slava* Antarctic whaling fleet. A proclamation of sovereignty in 1948 confirmed that His Majesty's Government in London had agreed to a transfer of sovereignty of the two islands to the

Government in Johannesburg. There has been a permanent South African presence, in the form of a meteorology and biology station, on Marion Island ever since.

A year later, South Africa sent further expeditions to Tristan da Cunha and Gough Island to exploit the islands' resources, and Marion and Gough Islands became sites for observatories of the IGY. South Africa established a presence in Antarctica proper on 8 January 1960, when she took over the Norwegian station, sited in Dronning Maud Land, from the Norsk Polar Institutt of Oslo.

Norway station (now named SANAE – South African National Antarctic Expedition) was subsequently replaced by a new base 30km (19mls) inland. SANAE was rebuilt in 1970/71, and again in 1978/79.

In 1969, four men wintered at Borgmassivet, a new inland camp 380km (236mls) south of SANAE, near the old Norwegian-British-Swedish expedition base of Maudheim. A new inland base, Grunehogna, was opened on 8 May 1971 in the Ahlmann Mountain Range 200km (124mls) south of SANAE. Five men overwintered there. Between 1971 and 1977, geology and geophysics parties wintered either at Borga or Grunehogna.

In 1978, the South African Survey vessel *Protea* started a programme of hydrographic

and oceanographic studies, especially on parameters pertaining to the movements of krill swarms, in the Scotia Sea.

There are persistent rumours that South Africa detonated a nuclear device just north of the Antarctic Treaty Zone, south-east of Bouveya Island in the South Atlantic, on 22 September 1979.

Current activities South Africa maintains two permanent sub-Antarctic stations, one permanent Antarctic mainland base and one summer-only station, although a second summer station, Sarie Marais, was opened in 1986/87. South Africa pays lip service to the notion of Norwegian sovereignty by annually requesting Norway's permission to maintain a presence in Dronning Maud Land. Biological/ oceanographic research programmes provide contributions to BIOMASS, whilst inland work is aimed at investigations of promising geological structures in Dronning Maud Land that might conceivably have mineral potential.

In 1986, South Africa announced plans to install an airstrip on Marion Island, which would take large transport planes. An environmental impact assessment recommended that the project should not proceed, and an announcement to that effect was made during May 1987.

THE U.K.

History The pioneering voyages of Cook (1772-75), Bransfield and Smith (1819-20), Weddell (1820-21 and 1822-23), John Biscoe (1830-32), Ross (1839-43) and the Challenger expeditions (1872-1876) all firmly established Britain at the forefront of 19th-century Antarctic exploration. The heroic expeditions of Borchgrevink (1898-1900 British Antarctic expedition), Scott (1902-04 and 1911-13), Bruce (1902-04), Shackleton (1907-09, 1914-16 and 1921-22) and Rymill (1934-37), the not so heroic British Imperial expedition (1920-22) and

the "Discovery" voyages (1924-51) carried the tradition well into the 20th century.

The UK registered the first claim to Antarctic Territory by a Letters Patent in 1908. The claim had to be adjusted by a further Letters Patent in 1917: the 1908 effort had, inadvertently, claimed part of the Argentine and Chilean Patagonia as well.

HALLEY STATION

This under-snow base has had to be replaced four times, as each of the successive structures has been crushed by the steadily shifting ice sheet (above and below).

The British claim included all Antarctic lands between 20° and 80°W to the Pole, together with the Scotia Arc Islands. The territory was to be administered as Falkland Islands Dependencies.

On the strength of this territorial claim, Britain began licensing the lucrative whaling industry in the South Atlantic and Southern Oceans, to the frustration of Argentina and Chile, both of whom had designs on the same territory. Neither country was to register its own claim until the 1940s. But the stage had been set for an increasingly bitter tripartite disagreement, which reached crisis proportions in the 1940s and 1950s.

During this period, all three countries increased their presence in the region and the installation of a new base by one country was always accompanied by formal protests from the other two. On 1 February 1952, tempers snapped and an Argentinian naval officer fired a machine-gun over the heads of a British party attempting to land supplies in Hope Bay. The incident highlighted the growing potential for conflict. Several

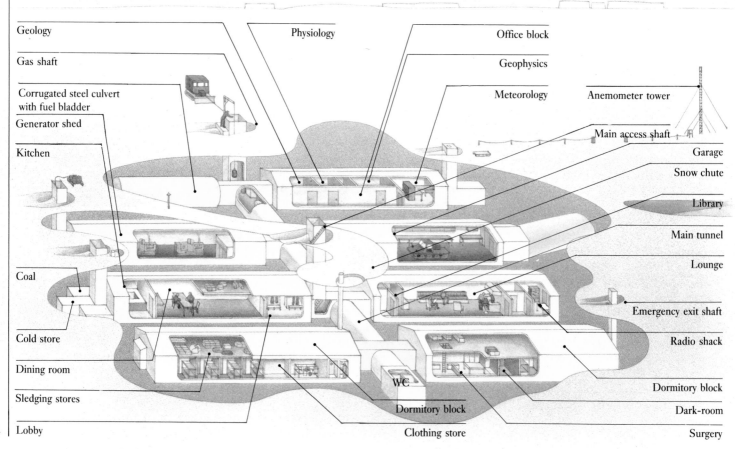

Geology
Gas shaft
Corrugated steel culvert with fuel bladder
Generator shed
Kitchen
Coal
Cold store
Dining room
Sledging stores
Lobby

Physiology

Office block
Geophysics
Meteorology
Anemometer tower
Main access shaft
Garage
Snow chute
Library
Main tunnel
Lounge
Emergency exit shaft
Radio shack
Dormitory block
Dark-room

WC
Dormitory block
Clothing store
Surgery

efforts to have the issue arbitrated in the International Court of Justice between 1950 and 1955 were unsuccessful.

Fortunately, the IGY released some of the pressure. The UK maintained up to 13 small bases or field huts during the 1956-58 period, of which Argentine Islands, Halley Bay and Port Lockroy made valuable contributions to the meteorological and atmospheric sciences programmes. Halley was established in 1956 by the Royal Geographical Society, and transferred to the Falkland Island Dependencies Survey in January 1959.

In the aftermath of the Antarctic Treaty, the UK introduced the name British Antarctic Territory for British possessions south of 60° latitude, whilst keeping the name Falkland Islands Dependencies for South Georgia and South Sandwich Islands. Several more bases opened and closed in the 1960s and 1970s.

The UK has maintained a total of 21 bases in British Antarctic Territory since World War II, although not all simultaneously. The programmes of the British Antarctic Survey are respected for the continuity of data sets, some dating back uninterrupted for several decades.

Current activities The British Antarctic Survey maintains five bases, extensive summer field operations encompassing nearly

BAS SCIENTIFIC STUDIES Research involves accurate distance measurement (above), and air and snow sampling (left).

2 million km² (1 million mls²) of territory, two research/supply vessels and three small aircraft, on an annual budget of about £14 million. The total summer population for all five UK bases is about 150 (including base personnel in transit), reducing to about 70 during the winter. Of the 70 over-winterers, about half are scientists.

In the aftermath of the Falklands/ Malvinas war, the British Antarctic Survey had its current annual budget more than doubled to enable it to maintain a higher presence in British Antarctic Territory. The Survey is planning to rebuild two bases, construct a hard airstrip at Rothera, and possibly purchase a new type of aircraft to extend field operations. A new air support system may even allow aircraft to fly into Rothera from the newly constructed Falklands/Malvinas airport.

These capital projects may cost in the order of £30 million, requiring extra Government funding at a time when the British Government appears ill-at-ease with its traditional Antarctic role. A condition for the doubling of the annual budget in 1982 was that there should be more emphasis on resources-orientated research, particularly in the field of mineral resources.

125

URUGUAY

History Uruguay's geopolitical attitude towards Antarctica, tempered by its historical role as a buffer state between Argentina and Brazil, has solidified in recent years. There is no doubt that Uruguay sees Antarctica as a strategic problem of immediate concern to her. Uruguay established an Antarctic Institute in 1968, and a commission for Antarctic studies in 1970. A request to join SCAR in 1983 was met with a polite refusal, and an explanation that full membership could only be considered after the establishment of a scientific research programme. When Uruguay announced its intentions to construct a station on King George Island, already the site of several stations, SCAR expressed concern that a new base on King George Island would not contribute significantly to Antarctic research, but might well contribute to environmental degradation. Uruguay acceded to the Treaty on 11 January 1980.

Uruguayan scientists did gain some Antarctic experience over several years with the Chilean research programme, and in 1981/82 two Uruguayan Air Force officers visited McMurdo station and the Amundsen-Scott South Pole station.

In November 1983, a Uruguayan army colonel studied the logistics of New Zealand's Scott and Vanda stations in preparation for Uruguay's first Antarctic expedition in December 1984.

Uruguay established its first research station in Collins Harbour, on the south coast of King George Island, despite SCAR's advice to go elsewhere. The new base, named Artigas, was installed with the assistance of a Chilean ship and helicopters, and inaugurated on 17 January 1985.

Current activities Artigas is capable of accommodating 12 to 15 people and, after serving as a summer station in 1984/85 and 1985/86, carried a complement of officers for the winter of 1987. Rudimentary programmes of meteorology, geomorphology and biology are conducted.

Uruguay obtained full Consultative Status within the Antarctic Treaty system, on 7 October 1985, thus creating an embarrassing procedural anomaly. Normally full Consultative Status presupposes a serious scientific contribution, usually under the auspices of SCAR, which Uruguay had not made. Its acceptance as a Consultative Member was, at least in part, a sop to Third World states in advance of the debate at the United Nations later in 1985.

THE U.S.A.

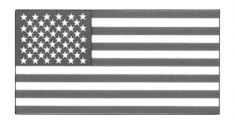

History American sealers were in the forefront of fur seal hunting expeditions into the Southern Ocean in the early part of the 19th century. Among them was Nathaniel Brown Palmer, who later claimed to have seen new lands south of the South Shetland Islands in November 1820. His discoveries were contemporaneous with those of the Russian, Bellingshausen, and the English sealer, William Smith. Commercial and national rivalries encouraged secrecy. It is difficult, if not impossible, to

SOUTH POLE DOME *HRH Prince Edward and a group of visitors descend into the American Amundsen Scott base, situated at the geographic pole.*

WINTER TRAVEL *Two scientists (top) return from work on the ice fields. With caterpillar tracks, this van (above) can move equipment equally well on snow or rock.*

ABBOT PEAK *A seismic recording station (right) has been set up on the slopes of volcanic Mount Erebus.*

determine who should be awarded the priority of discovery.

The pressure to locate new sealing and whaling grounds persuaded the United States to fund the US Exploring Expedition in 1838-42, which consisted of a six-vessel fleet under the overall command of Lieutenant Charles Wilkes. Wilkes made important discoveries along the Antarctic coastline that now bears his name.

America's Antarctic involvement in the 20th century is really the fruit of one man's personal obsession with polar regions. Richard Evelyn Byrd announced on 9 May 1926 that he had successfully flown over the North Pole. Three years later, on 29 November, Byrd repeated his achievement at the South Pole: the age of aviation had been well and truly introduced to Antarctica.

Byrd's 1928-30 Antarctic expedition was the first of several American expeditions between 1930 and 1947 (1928-30, 1933-35, 1935/36, 1938/39, 1939-41, 1946/47) responsible for mapping and photographing over 80 per cent of the continent. Under specific instruction from Congress, Byrd, Ellsworth and others were responsible for

claiming, or laying the basis for a claim to, over 2 million km^2 (1 million mls^2) on behalf of the American Government. The US never formally ratified such a claim, much of which would have been at the expense of existing claims. President Roosevelt's plan in 1939 permanently to occupy Antarctica as a basis for an American claim left no doubt about long-term ambitions.

Operation High-Jump, in 1946/47, was the largest Antarctic expedition ever, involving 4,700 servicemen, 51 scientists, 13 ships and 50 helicopters. High-Jump was intended to give the US experience in polar warfare operations, in preparation for a possible confrontation with the USSR across the Arctic basin. The subsequent emergence of Russian Antarctic interests created a deteriorating political situation that delayed the filing of an American claim. The timely organization of the IGY forestalled any immediate crisis.

America played an important role in the IGY. Like the USSR, the US took advantage of the "entente-scientifique" to secure a permanent foothold in strategically important areas of Antarctica, including the Ross Dependency (McMurdo station), Wilkes Land (Wilkes station), the Filchner Ice Shelf (Ellsworth station), Marie Byrd Land (Byrd station) and, most significantly, the South Pole itself (Scott-Amundsen base). Several temporary stations were opened to aid field science programmes, internal flights and expeditions. The US and New Zealand jointly operated Hallett station in

the Ross Sea area for several years, as an expression of the growing interdependence of the two Antarctic programmes.

In 1959, following the IGY, Wilkes station was transferred to the Australians, and Ellsworth to Argentina, whilst Little America V (the fifth in a succession of bases inaugurated by Byrd) was closed. The operation at McMurdo became enormously expanded, with a proliferation of buildings and support services, and a summer population of several hundred. By the mid-1970s, McMurdo was the largest base, capable of accommodating 800 men and women in summer, plus ships' crews.

Hallett station was closed in 1973 after 17 years of continuous operation. Plateau station (America's coldest, highest and most remote) operated between 1965 and 1969,

and Eights station, between 1963 and 1965. All five "Little Americas" floated out to sea on icebergs. In 1969, Siple station was built near the Ellsworth Mountains, less than 1,000km (621mls) from the Pole, and was completely rebuilt in 1979/80. The construction of Palmer station on Anvers Island, off the Antarctic Peninsular, completed in 1968, quickly promoted the Soviets to build a peninsular station, Bellingshausen, on King George Island. By 1974, the Palmer, Siple and the South Pole were entirely staffed by civilians.

Current activities With a budget of approximately US$120 million in 1986/87, the National Science Foundation (NSF) supported 70 research programmes of four permanent stations, three regular summer camps and a number of occasional summer sites and huts. In the latter category, the NSF normally opens a substantial summer site once every other year, as a site for field geology, geophysics, glaciology and terrestrial biology programmes. Some 40 to 60 scientists are supported at field camps between November and February.

In terms of geographical area covered, and the number of field staff supported, the US summer programme is the largest of any nation. There is a joint Anglo-American aeromagnetic survey of southern Palmer

CHAPEL OF THE SNOWS *This place of worship serves McMurdo base.*

SNOWED UP *An overnight blizzard can mean digging a way out in the morning.*

McMURDO STATION *An aerial view of the US base (below).*

INCONGRUOUS INTRUDERS *The awesome white landscape dwarfs the slimline sledge (above), while the beautiful sunset threatens to engulf McMurdo base (left).*

Land, Ellsworth Land and Marie Byrd Land currently in progress.

In total, in 1986/87, 33 meteorologists, 35 glaciologists and 53 geologists operated out of American summer camps. In total, the US Antarctic commitment in 1986/87 included over 1,500 personnel, of which 330 were scientists.

There are also comprehensive science programmes conducted at all winter bases. The winter population on American bases in 1986 was 150 men and 13 women, of which only 14 in total were scientists (five at the Pole, two at Palmer, five at McMurdo and two at Siple). This means that the US has the lowest proportion of wintering scientists to support staff of any major Antarctic nation (less that 12 per cent, compared to about 35-40 per cent for the UK, and a similar number for West Germany and France). Twenty-seven-year-old Dr Michelle Raney became the first woman to winter at the South Pole in 1979. In summer, the American programme more than makes up for its winter deficiency by supporting over 300 scientists.

THE U.S.S.R.

History Although there is some disagreement over who was the first man to set eyes on the Antarctic continent, one of the contenders for the honour is certainly the Russian explorer, Thaddeus von Bellingshausen, whose voyages in the Southern Ocean in 1819-21 were comparable in importance to those of Cook three decades earlier. Strangely, Russia showed no further interest in southern regions for 125 years.

In 1939, the UK warned Norway of a rumour circulating the diplomatic corridors of pre-war Europe that the USSR was planning an expedition to the Antarctic Peninsula. It was feared that the expedition might also have the intention of establishing Soviet sovereignty over Peter I Island, first discovered by Bellingshausen, but claimed by Norway in 1931.

Two weeks after Norway's 1939 proclamation of sovereignty over Dronning Maud Land, the USSR sent a note refusing to recognize Norway's claim and reserving judgement on the ownership of lands discovered by Russian expeditions. Such lands, however unspecified, included Dronning Maud Land, first approached by Bellingshausen on 27 January 1820.

The first practical Soviet step for a foothold in Antarctica began with a ratification of the International Whaling Commission on 20 November 1946. In December 1946, a Soviet whaling flotilla was despatched, consisting of a mother factory ship *Slava* and several catchers.

The *Slava* expeditions continued annually for two decades and, by 1960, included 16 whale catchers. The expeditions did not confine themselves to the commercial aspects of whaling but also carried scientists who carried out much valuable work in meteorology, sea-ice studies, biology and oceanography.

By 1950, it was clearly apparent to the Government of the US, which was investigating the political ramifications of making her own Antarctic claim, that the worsening political situation called for immediate and international attention.

Attempts to organize a conference involving the "sovereignty" countries of the United Kingdom, France, Norway, Australia, Argentina, Chile and New Zealand prompted a suspicious Soviet Government to state, on 8 June 1950, that "the Soviet Union reserves to itself the rights based on discoveries and explorations of Russian navigators and scientists, including the right to present corresponding territorial claims in Antarctica". The USSR would not countenance any attempt at an

HEAVY-DUTY VEHICLE *The rough terrain is no problem for this transporter.*

CLOSE-UP WORK *A Soviet mycologist examines specimens of fungi.*

international solution to Antarctica that did not include the USSR.

The statement, coming in the same year that the USSR developed its own nuclear capability, not only scuppered the proposed conference but also caused the US to reconsider the timing of its own potential claim. Antarctica was plunged into the frightening limbo of the Cold War, and did not emerge as a relatively settled issue until the Antarctic Treaty negotiations took place in 1959.

The Soviet contribution to the IGY was predictably prodigious: in their eyes, a whole continent was at stake. Of the 44 stations constructed in 1956-58, six were Soviet. The USSR utilized the political "cease-fire" of the IGY to construct a series of bases meant to cover as wide a geographical area as possible. Denied the opportunity of a base at the South Pole by the United States, the USSR constructed one at the Pole of Relative Inaccessibility (the point most distant from the sea).

At the termination of the IGY, a number of Russian bases were closed, but enough remained open to signal to the rest of the world the Soviet intention to remain in Antarctica come what may. In many respects, the continued Soviet presence acted as a spur to the successful negotiation of the Antarctic Treaty. New Russian stations soon replaced the moth-balled Pionerskaya, Komsomolskaya, Oaziz (transferred to Poland and renamed Dobrowolski) and Sovietskaya bases.

In 1959, Lazarevaskaya (on the Lazarev Ice Shelf) was opened, and in 1961 moved to the Schirmacher Hills, where it was renamed Novolazarevskaya. In 1962, Molodezhnaya was opened, and in 1968, Bellingshausen was established on King George Island. The main Soviet IGY base, Mirny, was totally rebuilt in 1973/74, and Vostok, Russia's prestigious inland base, was also refurbished. In 1980, a new base, Russkaya, was built with considerable difficulty on Cape Burks. Therefore, by the 1980s, the USSR had succeeded in building a ring of bases totally encircling the continent, representing the largest national operation in Antarctica. Although the Soviet Union does not recognize the validity of any Antarctic claim, it reserves the right to make one itself, and its strategically sited bases would provide a strong foundation in support of such a declaration.

Current activities In 1987, the Soviet Union maintained seven permanent winter stations, and up to six summer stations. A full range of atmospheric science, earth science and biological science programmes are supported. The total population in 1987 was 337, and the summer population often exceeds 700, not including ships' crews and officers, which would bring the total to well over 1,000 men and women. Over 5,000 Russians have wintered since 1956. The largest base, Molodezhnaya, has a winter population of 129. As well as base science, several field programmes presently concentrate on mineral resources evaluations.

The USSR also maintains comprehensive air support for its Antarctic operations and it is a stated Soviet ambition to be in a position to be able to airlift all Soviet Antarctic personnel into and out of Antarctica by 1990, thereby reducing the large number of ships presently required.

FAR FROM HOME *This makeshift, multiple signpost highlights the distance from family and friends.*

IMPACT OF BASES

Over the years, several attempts have been made to develop proper procedures for assessing and limiting the impact of scientific research activities on the Antarctic environment. Voluntary guidelines to limit pollution around Antarctic bases were adopted in 1972 and a code of environmental conduct was proposed in 1975.

The question of environmental impact assessment *per se* was not properly addressed until 1985 when SCAR presented a proposal recommending that any activities with a high probability of causing significant disruption should be subjected to rigorous investigation.

The report, entitled "Man's Impact on the Antarctic Environment", acknowledged that: "the majority of existing research stations were established in their current localities because these were the most convenient places for either logistical or scientific reasons and without thought for environmental effects". It also affirmed that there was often significant damage caused to the locality as a direct result of the stations being built.

Details of the SCAR report
The report went on to list some of the potential impacts that a permanent base could well have:

1 The actual construction may require a radical modification of the local habitat.

2 The generation of power and heat may result in the emission of gases, waste heat, dust and noise.

3 Oil spills and toxic wastes from shore-based laboratories may be released into the sea.

4 Supplying the base may require ships (with the potential for an accidental release of fuel oil in the pack-ice) or aircraft (which require sophisticated ground facilities, sometimes including hard-rock runways).

Wide-ranging pollution
The report pointed out that environmental impacts are not limited to the terrestrial ecosystem and living organisms. SCAR warned against: the "pollution" of the electro-magnetic spectrum by excessive radio-frequency radiation which interferes with scientific observations of electromagnetic phenomena; also the disturbance or removal of objects of scientific interest (fossils, meteorites, etc.) or damage to historic sites.

It also claimed that the biggest and most consistent problem is the disposal of waste. Solid wastes, such as discarded machinery, are offensive to the eye, it says, but of little biological significance. Plastic and rubber wastes carry longer-term consequences; small plastic particles are increasingly found in the digestive tracts of seabirds. SCAR recommended that such products "should, wherever possible, be removed from the area, but otherwise dumped on land. They should not be burnt unless they can be incinerated with adequate emission control."

Domestic and human wastes, kitchen refuse and natural wood can all be disposed of by placing in the sea with minimum treatment, SCAR recommended, but all radioactive waste, wastes containing high levels of heavy metals and harmful persistent organic compounds, should be removed from the Treaty area.

Proposals considered and shelved
Greenpeace and ASOC have proposed a series of recommendations on waste disposal. In a paper presented at the 1987 Consultative Meeting, it was suggested that the initial generation of waste materials should be minimized, that waste materials should be reused and recycled, and that, to the maximum

DUMPING AT SEA *Cost-effective but potentially damaging – scrapped machinery is dragged offshore and left on the pack-ice to sink when summer comes.*

ACCIDENTS WILL HAPPEN

When things go wrong, both the land and the atmosphere may be polluted, as can be seen from this garage fire (above). Very often, too little attention is paid to cleaning up the results. The skeleton of this crashed helicopter (right) lies in the Wright Valley long after the event.

HAZARD TO WILDLIFE *A junk heap on Anvers Island is an incongruous sight, as discarded machine parts and metal scrap are piled up and left. Such dumps create not only an eyesore, but a danger to wildlife.*

extent possible, remaining waste should be returned to the country of origin.

The SCAR report represented an important step forward, even though the responsibility for carrying out and making all the judgements on an Environmental Impact Assessment (EIA) lay with the state planning the activity. However, it was not translated into a Treaty recommendation or even into a suggested guideline. The issue was again considered at the 1987 Consultative meeting, but pending a further report from SCAR, it was again shelved.

The scale of the problem

Many bases in Antarctica have yet to solve the problems and practicalities of waste disposal. The US, for example, operates an open solid waste dump to service McMurdo base. Open burning is used to reduce the large volume of combustible materials and refuse is doused with waste fuel. During the 1983/84 Antarctic summer season 13,600–40,800l (1,000–3,000gal) of fuel were used for each burn, and the dump was burned five times that season.

The Australian Wilkes and Casey Stations' icy rubbish dumps were reported by SCAR member Dr Ron Lewis Smith in February, 1986. He commented that Wilkes Station appeared as it did when it was abandoned in 1969 in favour of Casey. Tinned and bottled food, machine parts, buildings, chemicals (including more than 200 boxes of tinned caustic soda spilling their contents on to the snow) metal drums, flares and even explosives were scattered over 1km^2 (0.4mls^2).

Casey's rubbish tip, covering 2,000m^2 (26,000ft^2) was a serious ecological hazard, he said. There was no separation of toxic materials from other non-combustible wastes. Skuas were found dead around the tip and scavenging birds have removed food scraps and dropped parts of them over a wide area.

Australia has since mounted a major clean-up effort at these bases.

Many bases also place solid wastes in metal drums and dispose of them at offshore dump sites. Larger pieces of machinery are often placed on the ice and left to sink with the summer thaw.

One such dump site, off McMurdo base, has been described by divers as "essentially dead, the sediment is so full of DFA [diesel fuel additive] it almost appears combustible! Clearly there was a massive spill of some sort and I doubt if that amount of DFA will be broken down in the near future ..."

Pollution can also travel to Antarctica from the rest of the world. A measurable increase of krypton-85 has been reported in the atmosphere since the advent of nuclear explosions and nuclear power plants. DDT has been found to occur in penguin fat and eggs.

NUKEY POO

The Antarctic Treaty bans nuclear explosions as well as the disposal of nuclear waste in the Antarctic but it does not forbid the use of nuclear power in the region.

In the 1960s the US installed an experimental 1.8MW pressurized-water reactor, nicknamed "Nukey Poo", at its McMurdo base to try to find a more economical way of providing heat and power.

It arrived by ship on 21 December 1961 and was sited half-way up Observation Hill near Mount Erebus, an active volcano. Power production began in July 1962 and, four years later, the US Navy claimed that it had broken the record for the longest continuous operation of a military nuclear reactor. In 1971 the power output was increased by 10 per cent.

During this whole period they claimed that the only serious problem had occurred in 1962 when hydrogen, a by-product from the reactor, caught fire but that there were no injuries or release of radioactivity.

In reality, the 10 years of the reactor's operation were an expensive story of shutdowns, fire damage and radiation leakages. In 1972, a temporary shutdown caused by coolant water leaking into the steam generator tank coincided with a Navy cost-effectiveness study. They concluded that it would be too expensive to overhaul and upgrade the plant, and, as a result, it was closed down and demolished over the next three Antarctic seasons at a cost of $1 million.

The reactor and 101 large drums of radioactive earth were shipped back to the US for burial. Later, another 11,000m^3 (14,500yds^3) of soil and rock were also removed and shipped back to the US. It took a further six years of cleaning up before the site was declared to be "decontaminated to levels as low as reasonably achievable", and it was finally released for unrestricted use in May 1979.

AN EXPENSIVE MISTAKE
US Navy personnel man the control room of the US nuclear power plant. The American plan to use nuclear energy in Antarctica to produce power economically for McMurdo base became a saga of accidents. It was finally shut down and removed, at great expense. Thousands of tonnes of contaminated soil and rock had to be shipped back to the United States.

THE FRENCH AIRSTRIP

One of the focal points of Greenpeace's Antarctic campaign has been its opposition to the French airstrip near the Dumont d'Urville base in Terre Adélie, due south of Tasmania, which it considers a breach of Antarctic Treaty rules.

The Dumont d'Urville base was constructed in the 1950s after the previous French base at Port Martin was destroyed by fire. Ironically, the site was chosen because of the richness of fauna and flora in the area, which made it almost unique in the whole of Antarctica, including as it does one of the very few emperor penguin colonies outside the Antarctic Peninsula.

The base has one major disadvantage, however. Ice conditions in the area only allow access by sea for two months of the year – January and February. This means that the French programme of research is limited compared to that of other countries.

So, in 1969 France began to think in terms of providing air support for the base. Over the following 10 years this was discussed several times by SCAR, but no specific proposals were put before the organization.

Blasting the site

However, in January 1983 construction work began on the site – a chain of five small islands in the Pointe Géologie Archipelago very close to the base. These islands lie in a straight line over a distance of about 1,100m (3,600ft). The plan was to blast the islands, levelling the surface and pushing the spoil into the sea to fill in the shallow channels between them. In this way an airstrip would be created, that was capable of handling the French Transall C160 aircraft, which would fly in from Hobart, Tasmania.

During that season and the next, explosives were used to move thousands of cubic metres of earth. At least 20 Adélie penguins and five Cape pigeons were killed and more than 1,500 Adélie eggs were destroyed.

The "Agreed Measures" of the Antarctic Treaty place explicit restrictions on human interference in relation

KILLED BY THE BLAST *The planned French airstrip crosses penguin colonies. This Adélie was killed by the rock blasting.*

to the Antarctic environment. In the construction of their airstrip, the French have been accused of breaching the Agreed Measures in several respects: killing native birds without a permit; failing to minimize harmful interference to the "normal living conditions of native mammals or birds"; and using explosives "close to concentrations of birds and seals".

Long-term harm

Greenpeace and other conservation groups have also opposed the project because of the likely long-term harm to the wildlife of the region. Since studies of the emperor penguin colony began, the number of couples has decreased from 6,500 to about 3,500, due in part to human presence. The airstrip will cut off the emperors' favoured access to their breeding colony.

Many other bird species, including 2,500-3,000 pairs of Adélie penguins, 300 pairs of Cape pigeons, 100 pairs of snow petrels and 100 pairs of Wilson's storm petrels will also be directly harmed because construction will destroy their nesting areas.

Inadequate report

Greenpeace was alerted to the project by a contact within the French Ministry

information was made public by environmental groups in January 1983. Shortly afterwards France published an inadequate Environmental Impact Report on the project, which drew severe criticism from many scientists. (The Code of Conduct for Antarctic Expeditions and Station Activities suggests that environmental impact statements should be prepared well in advance of any work taking place.)

On 21 March 1983 the French Academy of Sciences passed a strongly worded resolution against the project, and two French biologists publicly criticized it in an open letter to *Le Monde*.

In December 1983 Greenpeace acquired photographs and film of the construction activities of the previous season, including images of large explosions and dead penguins. This evidence was presented to the Antarctic Treaty Consultative Parties (ATCPs) at their January 1984 minerals meeting in Washington DC but no action was taken by them.

BLASTING A PATH
Explosives were used in an attempt to connect the islands of the Pointe Géologie Archipelago, so that an airstrip could be built there.

International reassessment

Continued pressure on the French Government led to the appointment of a *Comite des Sages* (known as the Thaler Committee) in the early part of 1984 to re-examine the environmental impact report and to make recommendations to the French Government. Its eight members included two foreign scientists, thus recognizing for the first time that the issue was of international importance.

The Committee recommended abandoning the project and considering other less harmful options, and the undertaking of a new impact study. It concluded that it "hoped to see the project abandoned ..."

These conclusions were kept secret for many months for political reasons. On 29 June 1984, six Greenpeace demonstrators, dressed as emperor penguins, scaled the façade of the *Terres Australes et Antarctiques Françaises* (TAAF) building in Paris to demand the release of the report. Several were arrested and brutally treated by police. TAAF, the promoters of the project, finally released it in September.

The following month this was followed by a second Environmental Impact Assessment, covering many of the criticisms of the first, but still evaluating alternative plans only in terms of logistical advantages or disadvantages not, as suggested by the Code of Conduct, environmental effects.

This new EIA concluded that the project should proceed, although it acknowledged that "there is no avoiding the fact that birds' nests and eggs will be lost and some adult birds possibly killed outright in the blasting for construction".

Its estimate of an average 10 per cent drop in fertility across eight bird species was based on dubious data.

Protests and resolutions

On 22 October 1984, Greenpeace protesters boarded the French-chartered vessel, *Polarbjorn*, in Le Havre and occupied the mast for 56 hours in protest. They claimed that it was carrying equipment for the construction of the projected airstrip.

In November 1984, members of the prestigious International Union for the Conservation of Nature and Natural Resources (IUCN) passed a resolution which called upon the French Government to consider other options. The resolution requested that they and the Antarctic Treaty Consultative Parties "thoroughly study the option of protecting Pointe Géologie from further construction".

In December 1984, Greenpeace conducted a further action against the *Polarbjorn* in Hobart, *en route* to the Antarctic, involving *Vega* and six other boats. Demonstrations were also held in Adelaide, Sydney and Melbourne.

Construction to continue

In October 1985 the French delegation at the 13th Antarctic Treaty Consultative Meeting announced that the construction of the airstrip would proceed but did not indicate when.

The ATCPs have exhibited little will to deal with the problem. At least one government – New Zealand – is known to have made direct representations to France over the issue, and the Australian Government has reportedly made a formal determination that a breach of the Agreed Measures has occurred. But the question has never been formally raised in the context of an Antarctic Treaty Consultative Meeting or in any other meetings of the Treaty System. As a result, the French have never formally been asked to explain the alleged violations.

More alarmingly, in November 1987 Bernard Pons, the French Minister of Overseas Partments and Territories, announced that work on the 1,000m (3,300ft) runway at Terre Adélie, which was interrupted in 1984, would resume immediately and would last for about five years at a cost of 100 million francs (£10 million).

IMPACT OF TOURISM

Permanent human settlements and scientific research aside, the other human incursion in Antarctica is tourism, the only commercial activity on the continent itself and one which is proving increasingly popular.

The earliest tourist flight to Antarctica was from Chile in 1956 but it was not until 1977 that regular flights were instituted by Qantas and Air New Zealand using 747s and DC10s. Over the next two seasons, some 11,000 passengers overflew the continent, an 11-hour journey of which some 90 minutes were spent over Antarctica.

Such flights ceased abruptly after an Air New Zealand DC10 crashed into Mount Erebus on 28 November 1979, killing all 257 passengers and crew.

Antarctic cruises

Cruise liners have been operating in the area since 1958 and, in the period up to 1980, more than 80 voyages have been undertaken by passenger liners from Argentina, Chile, USA, Spain and West Germany, carrying an estimated total of 17,000 passengers.

By far the largest number of cruises have been arranged by two US-based firms – Lindblad Travel and Society Expeditions. They have usually travelled from the tip of South America, down the Antarctic Peninsula, and sometimes round to the Ross Sea area, making landfalls at research bases, historical sites and penguin rookeries.

Chile in particular has recently been expanding its tourist activities in the region, flying tourists into King George Island where a hotel, bank and supermarket are being built.

Independent expeditions

Tourism, in the jargon of the Antarctic Treaty System, also includes non-governmental expeditions. In the last few years these have included:

1 Project Blizzard, a private Australian expedition which, in 1984/85, aimed to

JUST PASSING THROUGH *A whalers' oil tank, now disused, forms the backdrop to a group of tourists visiting Deception Island.*

restore the remains of Sir Douglas Mawson's hut at Cape Denison. A second expedition completed the restoration work the following year.

2 In the Footsteps of Scott, a private expedition, which, at the end of 1985, successfully retraced on foot the route taken by Scott from Cape Evans to the South Pole in 1911-1912. Literally minutes after their arrival at the Pole, the three walkers learnt that their supply ship, *Southern Quest*, had been crushed by ice in the Ross Sea and had sunk.

3 In 1987, a private Norwegian expedition, which aimed to retrace Roald Amundsen's route from the Ross Sea to the South Pole. They abandoned their journey 386km (240mls) from the Pole.

Disruption to bases

The question of the effects of Antarctic tourism has been addressed several times at meetings of Treaty members. Under the various rulings and guidelines, it is the responsibility of each Consultative Party to ensure that any of its nationals who are part of a tourist or non-governmental expedition abide by the Agreed Measures. However, it is both politically unacceptable and legally difficult for any Consultative Party to prevent any non-governmental expedition from exploring or traversing the frozen continent.

At least seven commercial tourist voyages and a number of smaller expeditions have got into difficulties and required assistance. ATCPs have expressed worries that tourism raises the possibility of expensive disruptions to personnel at research stations and a hazard to life and, in some instances, base equipment.

When the Air New Zealand DC10 crashed on Mount Erebus, 20 hours of US Hercules C130 flying time and many hours of helicopter time were diverted from the tightly scheduled scientific programmes. In January 1968 the tourist ship *Magga Dan* ran aground at McMurdo for several days and the *Lindblad Explorer* has twice run aground in the Antarctic Peninsula, requiring expensive and disruptive rescue and repair operations to be mounted.

Impact on the natural environment
Aside from the disruption caused to bases, by an influx of tourists or by distress calls, there is the question of the additional pressure tourists place on the natural environment. Fragile vegetation could easily be destroyed, and nesting and breeding grounds disrupted. Tourists could unwittingly spread bird or plant diseases and introduce new kinds of organisms to the Antarctic. The expansion of tourist facilities may also have a significant impact. At present this is a small problem, but one that is rapidly worsening. Under a World Park regime there would have to be regulation of some sort.

CAPE ROYDS
The Cape Royds Adélie penguin colony is the most southerly known penguin rookery, situated about 32km (20mls) from the New Zealand and US bases on Ross Island.

From 1956 onwards, numbers of penguins began to decline steeply. This was owing to the steady flow of base personnel, as well as congressmen, parliamentarians, journalists, diplomats, soldiers and sailors, who were visiting Antarctica as the guests of the two countries' governments.

In the summer of 1961, helicopter flights were recorded on almost every fine day. Each landing scattered the penguins, breaking the breeding routine, exposing eggs and chicks to skua predation, and unsettling young birds seeking nest sites.

The situation was aggravated by the additional attraction to visitors of Shackleton's hut, close to the colony. It was clear that such interference was leading to reduced breeding success among the penguins and the colony is now a protected site.

OCEAN VOYAGES
Tourists can now see the harsh beauty of Antarctica in comfort. The Lindblad Explorer, *seen here off Cape Hallet, has become a regular visitor, bringing groups on sight-seeing trips.*

A special attraction on board is a series of lectures by staff who are experts in such fields as art, ornithology and oceanography. These lectures include an historical account of Antarctica and man's presence there.

In addition to the ocean voyages, tourists are now flown to the South Pole, although this is, of course, rather expensive.

PROTECTED SITES

The near pristine and fragile nature of Antarctica's environment demands special treatment. Not only does it contain unique life forms but, as we have seen, its relatively untouched nature provides us with invaluable baseline information for understanding complex and diverse planetary systems.

The first attempt to preserve Antarctica's environment came in 1960 at a meeting of SCAR at which a series of general rules of conduct for expeditions operating in Antarctica were drawn up. Four years later this led to a more elaborate set of Agreed Measures for the Conservation of Antarctic Flora and Fauna. Under this ruling the whole area south of 60°S was deemed to be a "special conservation area".

Agreed Measures
The four main points of the Agreed Measures were as follows:

1) The killing, wounding or capturing of any native mammal (excluding whales, which were dealt with under the International Whaling Convention) or bird is prohibited. Special permits can be issued to provide indispensable food for man or dogs, to provide specimens for scientific study or information, and, in addition, to provide specimens for museums or zoos.

2) Appropriate measures must be taken by participating governments to ensure minimal harmful interference with the normal living conditions of native mammals and birds and to alleviate pollution of coastal waters.

(Examples of "harmful interference" include flying helicopters or driving

WILDLIFE STUDIES
A naturalist examines a one-day old blue-eyed shag on Signy Island.

vehicles too close to bird and seal colonies, disturbing such colonies in any way during the breeding season, and using explosives or firearms in close proximity to such colonies.)

3) The introduction of any non-indigenous species, parasites and diseases is forbidden. The only exceptions are sledge dogs, domestic animals and plants, laboratory animals and plants, including viruses, bacteria, yeasts and fungi – all of which have to be kept under carefully controlled conditions, and require a permit.

4) A network of "Specially Protected Areas" (SPAs), considered to be "unique natural ecological systems" or of "outstanding interest", should be established straight away.

This final measure was supplemented by the creation of a network of Sites of Special Scientific Interest (SSSIs), beginning in 1975.

SPAs, SSSIs and SMAs
There are some profound differences between the concepts of SPAs and SSSIs. The designation of an area as an SPA includes restrictions on entry into, and activities within, the site; an SSSI designation is meant to safeguard scientific research on the site, rather than the site itself. Restrictions are more relaxed for SSSIs, and the designation is reviewed periodically.

Proposals to create marine SPAs have consistently been blocked by the Soviet Union, ostensibly on the grounds that such a designation might interfere with the rights of states on the high seas.

More recently Australia has proposed the creation of Specially Managed Areas (SMAs), an intermediate status between the protection afforded by SSSIs and SPAs, and the absence of

any protection at all. This designation could be applied to large areas of Antarctica in order to allow scientific research but forbid more harmful activities to take place there.

Bending the rules

Such measures are presented as an ambitious experiment in conservation. Yet there is no express prohibition on the erection of a base within an SPA and, if there is a conflict between a base

siting and an area proposed as an SPA, the problem has usually been solved by "adapting" or terminating the SPA.

The best example of this occurred at the Fildes Peninsula SPA where the Soviet Union and Chile both set up bases in 1968. Their mere presence was bound to have a damaging effect.

The other Consultative Parties, instead of raising objections at that year's Consultative Meeting, actually reduced the SPA to a fraction of its

former size in order to exclude the Soviet Bellingshausen station and surrounding areas in which travel and disturbance were inevitable.

The original SPA designation centred around several small lakes of outstanding ecological interest. As a result of the 1968 decision, only the most interesting one retained its protected status. This reduced area was so affected by the presence of bases that its SPA status was terminated in 1975.

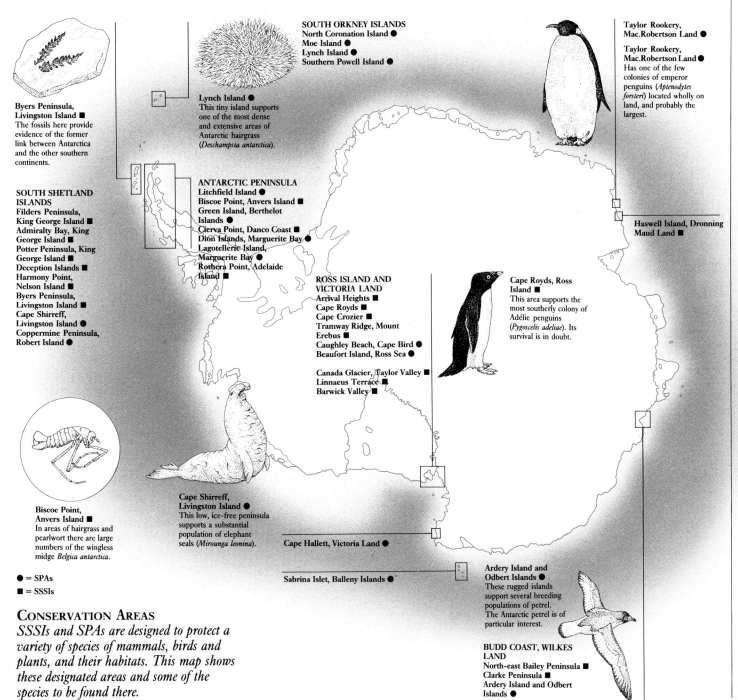

Byers Peninsula, Livingston Island ■
The fossils here provide evidence of the former link between Antarctica and the other southern continents.

SOUTH SHETLAND ISLANDS
Filders Peninsula, King George Island ■
Admiralty Bay, King George Island ■
Potter Peninsula, King George Island ■
Deception Islands ■
Harmony Point, Nelson Island ●
Byers Peninsula, Livingston Island ■
Cape Shirreff, Livingston Island ●
Coppermine Peninsula, Robert Island ●

SOUTH ORKNEY ISLANDS
North Coronation Island ●
Moe Island ●
Lynch Island ●
Southern Powell Island ●

Lynch Island ●
This tiny island supports one of the most dense and extensive areas of Antarctic hairgrass (*Deschampsia antarctica*).

ANTARCTIC PENINSULA
Litchfield Island ●
Biscoe Point, Anvers Island ■
Green Island, Berthelot Islands ●
Cierva Point, Danco Coast ■
Dion Islands, Marguerite Bay ●
Lagotellerie Island, Marguerite Bay ●
Rothera Point, Adelaide Island ■

ROSS ISLAND AND VICTORIA LAND
Arrival Heights ■
Cape Royds ■
Cape Crozier ■
Tramway Ridge, Mount Erebus ■
Caughley Beach, Cape Bird ●
Beaufort Island, Ross Sea ●

Canada Glacier, Taylor Valley ■
Linnaeus Terrace ■
Barwick Valley ■

Cape Royds, Ross Island ■
This area supports the most southerly colony of Adélie penguins (*Pygoscelis adeliae*). Its survival is in doubt.

Taylor Rookery, Mac.Robertson Land ●

Taylor Rookery, Mac.Robertson Land ●
Has one of the few colonies of emperor penguins (*Aptenodytes forsteri*) located wholly on land, and probably the largest.

Haswell Island, Dronning Maud Land ■

Biscoe Point, Anvers Island ■
In areas of hairgrass and pearlwort there are large numbers of the wingless midge *Belgica antarctica*.

● = SPAs
■ = SSSIs

Cape Shirreff, Livingston Island ●
This low, ice-free peninsula supports a substantial population of elephant seals (*Mirounga leonina*).

Cape Hallett, Victoria Land ●

Sabrina Islet, Balleny Islands ●

Ardery Island and Odbert Islands ●
These rugged islands support several breeding populations of petrel. The Antarctic petrel is of particular interest.

BUDD COAST, WILKES LAND
North-east Bailey Peninsula ■
Clarke Peninsula ■
Ardery Island and Odbert Islands ●

CONSERVATION AREAS
SSSIs and SPAs are designed to protect a variety of species of mammals, birds and plants, and their habitats. This map shows these designated areas and some of the species to be found there.

141

SEALING

Captain Cook's accounts of vast seal populations led British and American sealers south and, beginning in 1784, they systematically butchered their way through the fur seal breeding grounds until the animals were almost extinct. The fur was used for making slippers and little else.

Sealers plundered the Southern Ocean for fur seals throughout the 1900s, stopping only when stocks were virtually exhausted. They then turned to the elephant seals for their blubber. However, this was a less profitable operation, and by the beginning of the 20th century, the old-fashioned sealing operations had virtually disappeared.

In 1910 the British re-established the killing of elephant seals on South Georgia under a licensing system. A maximum kill of 6,000 adult bulls per year was allowed and this operation continued profitably, in more or less the same form, until 1964.

Gradual fur seal recovery

The fur seals took longer than other species to recover from the wholesale slaughter. After the last catch in South Georgia (1907) no fur seals were seen apart from solitary sightings, until 12 pups were found on Bird Island in the 1930s. By 1956 their numbers had increased to an estimated 3,500 and are still increasing at a rate of around 16 per cent a year. The species population as a whole is now thought to be back to its original numbers.

Limiting commercial exploration

Crabeater, Weddell, Ross and leopard seals have never been the subject of commercial slaughter in modern times except for two occasions. A 1964 expedition by the Norwegian sealer, *Polarhar*, took 1,127 seals but found the hunt too difficult and uneconomic to make it worth continuing. A 1972 Soviet expedition took 1,000 seals and came to similar conclusions.

That same year the Treaty states formulated the Convention for the

Conservation of Antarctic Seals (CCAS), which was formally ratified on 11 March 1978.

This called for the total protection of the fur, elephant and Ross seals south of latitude 60°S and set annual catch limits on the crabeater of 175,000, the leopard, 12,000 and the Weddell seals, 5,000. It stipulated that all killing must be done humanely and quickly, that a proper record must be kept of all animals killed and that they must not be hunted in open water, unless they are being taken for the purposes of scientific research.

Seasons for sealing

The Convention also established a closed season from 1 March to 31 August, and created six sealing zones (each of which is closed to sealing for a year in rotation). Three reserves were designated in which it is forbidden to capture or kill seals.

PROTECTION OF SEALS *The killing of seals for their fur and blubber devastated many Antarctic species. Now international measures have been taken to protect them, and numbers appear to be recovering.*

Most recently, during the 1986/87 Antarctic summer season, the Soviets killed 4,804 seals in the Indian/Pacific sector of the Southern Ocean. The majority killed were crabeaters (4,014) with smaller numbers of leopard seals (649) and Weddell seals (109). A few Ross seals and southern elephant seals were also taken.

It is legally permissible to take any number of any species in the name of scientific research, as long as the "objectives and principles" of CCAS are complied with. It remains to be seen whether this catch is considered to be a meaningful research project aimed at gaining a better understanding of the marine ecosystem.

WHALING

For more than 60 years, the Southern Ocean was the world's major whaling ground. The whaling industry was so intensive that the blue whale population now stands at less than one per cent of its original level.

The first whaling station was established in the region in 1904 by a Norwegian company who, in their first season, with one factory and a single catcher boat, took 195 whales.

The success of this operation led to an explosive growth in Antarctic whaling. By 1912/13 there were six land stations, 21 factory ships and 62 catcher boats, which killed and processed 10,760 whales. By 1930/31, the seasonal kill had risen to 40,000 and, for the next 20 years, except during the war, the slaughter continued at this level.

Targetted whales

Initially, humpback whales were the main target as they swam close to land but, as their numbers dwindled and whaling ships became able to operate further from port facilities, attention focused on the blue whale, which remained the preferred species until their numbers also declined.

In the 1950s and during the early 1960s, whaling operations centred on the fin whale. During the latter period, the Antarctic fleets were also catching at least 4,000 sperm whales a year in the Southern Ocean. The International Whaling Commission (IWC) did not set any limits on sperm whaling until as late as 1970.

The sei whale became the target species in the late 1960s and into the 1970s, but fin and sei whaling were closed in 1976 and 1978 respectively. By then the minke had become the most hunted whale species.

Catches of orcas have generally always been small, although the Soviets did take 916 in 1980; orcas fall under the category of small cetaceans, which are outside IWC regulations.

The end of commercial whaling

The IWC was established in 1946 to regulate "the orderly development of the whaling industry". Unfortunately, it was widely perceived as failing in this task, and from the early 1970s onwards conservation groups began an intense international struggle aimed at saving whale stocks from further depletion.

It is estimated that the population of blue whales is now less than 1 per cent of its original level; humpbacks are down to 3 per cent, and fin whales 20 per cent. All three species are now afforded complete protection, as are sei and sperm whales.

Since 1969, the only whaling fleets operating in Antarctica have been the those of the Soviets and Japan. The Soviets have since stopped. Conservationist pressure led to the IWC agreeing, in 1982, to a whaling moratorium that came into effect in 1986. It will be reviewed in 1990 after an assessment of whale stocks worldwide.

The beginning of scientific whaling

"Scientific whaling" for so-called research purposes is now the controversial issue, a loophole in the regulations that many whalers are exploiting.

Japan used the scientific research approach at the 1987 IWC meeting when it put forward a proposal calling for 825 minke whales and 50 sperm whales to be taken from the Antarctic annually for 12 years. This plan was criticized by the Scientific Committee of the IWC, chiefly because of the infeasibility of the research methods that were proposed.

Environmentalists filed suits in the US District Court, asking that fishery sanctions be imposed against the Japanese, arguing that the hunt was really a continuation of commercial whaling. The US Government told Japan that it would be subject to sanctions if it went forward with the proposal.

Japan then submitted a revised research programme for 300 minke whales and asked that this be reviewed by the Scientific Committee of the IWC. Lobbying and counter-lobbying over the issue of "scientific whaling" looks set to continue in this manner for some considerable time.

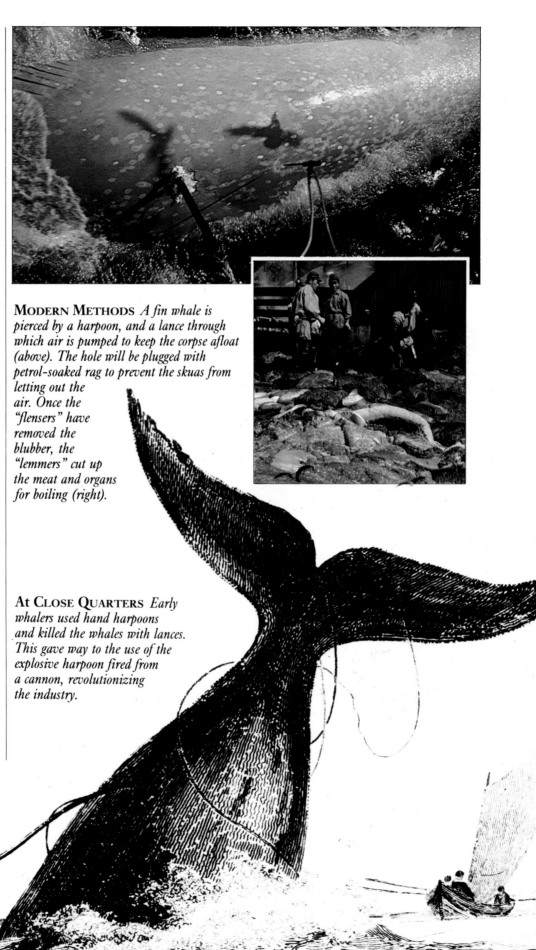

MODERN METHODS *A fin whale is pierced by a harpoon, and a lance through which air is pumped to keep the corpse afloat (above). The hole will be plugged with petrol-soaked rag to prevent the skuas from letting out the air. Once the "flensers" have removed the blubber, the "lemmers" cut up the meat and organs for boiling (right).*

At Close Quarters *Early whalers used hand harpoons and killed the whales with lances. This gave way to the use of the explosive harpoon fired from a cannon, revolutionizing the industry.*

FISHING

Large-scale trawling for fin fish began in the 1960s around South Georgia and Kerguelen, and increased with enormous rapidity. In 1970/71 the fleets of the USSR, East Germany and Poland caught over 400,000 tonnes of fish in the one season.

Antarctic fish are especially vulnerable to over-fishing because most species take a long time to mature. Juvenile fish and adults are found in the same fishing grounds and many undersize fish are caught in fisheries where there are no mesh-size regulations.

Since 1969, three of the four most important commercial fin fish species – marbled notothenia, scaled notothenia and Patagonian toothfish – have been very heavily over-fished in the area regulated by the Convention on the Conservation of Marine Living Resources, and few controls have been introduced. As for the fourth species, the Antarctic ice-fish, there are grave fears for the future of the stock after a catch of 162,598 tonnes in 1982/83.

OVER-FISHING *The catch from a bottom trawl net off the South Orkneys is almost entirely ice-fish. Very large catches in recent years raise fears of over-fishing.*

HARVESTING KRILL

The first exploratory fishing for krill was carried out by the Soviets and Japanese in the 1960s. It was slow to become established, owing to the remoteness of the fishing grounds from major ports and the fact that krill was a new resource, and required new catching and processing technology to be developed. In addition, new markets had to be established.

The fishing season for krill lasts for three to five months, with the densest swarms found during the period from January to April. The swarming habit of krill led to the use of echo-location equipment and a mid-water trawl, which requires powerful vessels because of the high drag of the small-meshed nets.

The total krill catch steadily increased to a peak in 1981/82 of 528,000 tonnes. It declined steadily to 128,000 tonnes in the 1983/84 season, and subsequently increased again to 436,445 tonnes in 1985/86, dropping to 376,527 tonnes in 1986/87. Japan, Poland and the USSR are the main krill fishing nations at present but five others have been involved.

ANTARCTIC KRILL
DISTRIBUTION

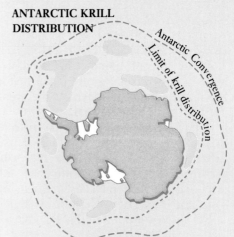

A good source of protein
In the live state, krill contain a maximum of 16 per cent protein and 7 per cent fats; in a processed form, protein content can be as high as 50 per cent. Krill also contain significant amounts of A, D and B group vitamins and are

rich in calcium, copper, iron, magnesium and phosphorus.

Most of the present krill catch is used as animal feed, a practice carried out on a commercial scale in Russia and Japan for years, even though in many cases it remains more expensive than fish meal. In a Polish experiment, cows fed on krill had a high incidence of abortion.

There was euphoric speculation in the 1960s and 1970s that krill might provide an answer to chronic protein shortages in developing countries, but it has now been found that krill live much longer and are less productive than previously thought. The discovery

Upsetting the Balance
Already being sold in Japan and other countries, krill is the subject of increasing commercial interest, which may threaten the entire Antarctic ecosystem.

南極 オキアミ
OKIAMI
食指を動かせ、南氷洋。

きれいな南氷洋からやって光た
80年代の高タンパク源！
なにするものぞ200カイリ！

アイデア生かして、いろいろ
オキアミ料理15選。

that krill "swarms" are not that predictable has also discredited the idea of vast, sustainable catches.

Only the industrialized nations possess the capital and technology to exploit krill and they have proved to be the main market for krill products, including its use for human consumption. In Japan, whole krill achieved satisfactory sales, as did a krill cheese spread ("Koral") in the Soviet Union. Chile produces frozen krill "sticks" like fish fingers, and the Norwegians used it in soup products.

Fit for human consumption?
In the early days scientific research suggested that krill as a human food was suspect due to fluoride levels seven to 24 times higher than those permitted by the US Food and Drug Administration. The fluoride was found to be a product of the decomposition that begins within a few hours of the krill being caught. They must be frozen rapidly to below $-20°C$ ($-4°F$) and then stored at a similar temperature to remain fit for consumption.

The technical problem of achieving this now appears to have been solved and interest in krill harvesting has picked up as a result.

The new "super trawlers"
One indication of the new interest is the pressure of a new class of Soviet "super trawler" in Antarctic waters. The first of those to go into operation was the *More Lazareva*, a 114m (370ft), multi-purpose trawler, weighing 6,392 tonnes, designed to operate in all weather conditions. It carries on board a fully automatic processing plant designed to freeze and can fish, crab and krill. The krill can be processed and canned at the rate of 150kg (330lb) or 1,500 cans an hour.

If no limits are set, krill catches could increase rapidly. This could lead potentially to disastrous disturbances throughout the entire ecosystem, since we do not know enough about krill to be able to predict the long-term effects of high catch rates.

MARINE CONSERVATION

In an effort to find a means of regulating existing fishery operations in general and the threat of an expanded krill fishery in particular, the Convention on the Conservation of Marine Living Resources (CCAMLR) was concluded in May 1980 and brought into effect in April 1982.

The CCAMLR represented a major breakthrough in marine conservation because, instead of considering each species separately, it provided for the effective conservation of the Antarctic ecosystem as a whole.

The Convention's objective was the "maintenance of the ecological relationships between harvested, dependent and related populations of Antarctic marine living resources".

The concept of CCAMLR represented a victory for conservation interests but those same interests are now highly critical of the way in which the Convention is being administered.

Members of the Convention

In order to deal with "outsider" fishing nations like South Korea, who were neither signatories nor Consultative Parties to the Antarctic Treaty, the Convention was set up as a free-standing agreement in its own right, which any state can join. Further, they can become members of the decision-making Commission if in fact they are "engaged in research or harvesting activities in relation to marine living resources in the Antarctic".

However, the Treaty powers have special status, automatically becoming members of the new Commission even if not engaged in harvesting or research.

Powers of the Convention

In most of the world's oceans, nation states restrict fishing off their shores and impose a catch quota on both national and foreign fishing fleets. Owing to the international nature of the Southern Ocean, this management function is assigned to the Convention's Commission, for example, in November

THE BIOMASS EMBLEM

1987, the Commission assigned a total catch limit of 35,000 tonnes on *Champsocephalus gunnari*, set up a catch-reporting system for the species, and imposed a closed season.

Voting under the Convention is by consensus on matters of substance, rather than by a simple majority. Thus, any single country has the power of "veto". Serious conservation measures have been consistently blocked by the fishing states – mainly the USSR. All that the other Commission members can do to make a country comply is to apply political pressure.

Even when measures are accepted, there is again no mechanism for enforcement; moreover, a country can renege on agreed measures within 90 days of accepting them.

Scientific Committee

The key to the successful operation of such a Convention is scientific data, and the task of modelling and understanding more fully the Southern Ocean ecosystem was placed in the hands of a scientific committee.

The scientific prospects were very exciting. If the relatively simple Antarctic ecosystem could be modelled and managed, in order to understand the effects of exploiting several species at different levels in the food chain, it might shed light on how to manage more complex fisheries.

Enforcement difficulties

One of the first requirements was detailed information on annual fish and krill catches, a highly sensitive subject for the fishing states, who have systematically failed to provide catch data in time for the meetings of the CCAMLR Scientific Committee, even though this is one of the requirements of the Convention. Nor have they provided complete catch data for the years they were fishing the Southern Ocean prior to the Convention coming into effect.

Annual catch data is supplied to the UN Food and Agricultural Organization (FAO) but this is very general and does not list what catches are made in which areas. It also fails to satisfy the requirements of CCAMLR's ecosystem objectives, which call for information on "dependent and related populations". Such detail is necessary to discover whether krill fishing is interfering with feeding whales and thus endangering their recovery.

As a result, an international information-gathering operation known as BIOMASS was set in motion.

The aims of BIOMASS

BIOMASS is an acronym for the Biological Investigation of Marine Antarctic Systems and Stocks. Its officially stated aim was "to gain a deeper understanding of the structure and the dynamic functioning of the

Antarctic marine ecosystem as a basis for the future management of potential living resources".

Instituted by SCAR, it was organized in two main stages, which are known as FIBEX and SIBEX – First and Second International Biological Experiments.

The findings of FIBEX and SIBEX

FIBEX was undertaken in the austral spring of 1981 and involved 17 ships from 10 nations who correlated weather and ocean information with observations of animals and krill.

Its most spectacular find was made by the US vessel RV *Melville*, which sighted a "superswarm" of krill near Elephant Island. Estimated at 10 million tonnes, the swarm occupied several square kilometres of sea to a depth of 180m (600ft) below the surface.

Scientists estimated that this single school was equal to about one-seventh of the world's total fish and shellfish catch for a year. News of this unexpected and so-far unrepeated discovery brought 35 Soviet fishing vessels to the area in order to reap the benefits.

SIBEX phase 1 followed in 1983/84 and phase 2 in 1984/85. Again, multidimensional observations were taken, ranging from the temperature, salinity and oxygen concentration of sea water to weather conditions, visibility and ice cover; and from the concentrations of krill and phytoplankton to sightings of whales, seals and seabirds.

Storage of data

All the data from these massive information-gathering operations is stored on a Honeywell computer at the BIOMASS data centre at Cambridge, UK. All groups of data are stored with the date and time the observations and measurements were made. Thus, bird observations can be correlated with krill swarms, and net samples can be correlated with acoustic data.

The data is also categorized by the name of the ship which supplied it, making it easy to examine the results from a single country or those from a group of countries.

A fuller picture

Such information-gathering exercises are to be welcomed if we are to gain a fuller picture of the Southern Ocean ecosystem; however, some scientists who participated in BIOMASS are disappointed that hoped-for levels of international cooperation were not achieved.

FIBEX (1980/81)	SIBEX I (1983/84)	SIBEX II (1984/85)
Walther Herwig (FRG)	Professor Siedlecki (POL)	Almirante Irizar (ARG)
Professor Siedlecki (POL)	S.A. Agulhas (SA)	Professor W. Besnard (BRA)
Marion Dufresne (FR)	Almirante Irizar (ARG)	R/V Alcazar (CHILE)
Nella Dan (AUS)	R/V Alcazar (CHILE)	Polarstern (FRG)
R/V Melville (US)	Polarstern (FRG)	Walther Herwig (FRG)
A.R.A. Homburg (ARG)	Nella Dan (AUS)	R.R.S. John Biscoe (UK)
M/N Itzumi (CHILE)	Kaiyo Maru (JAP)	Nella Dan (AUS)
S.A. Agulhas (SA)	S.A. Africana II (SA)	Marion Dufresne (FR)
Odessey (USSR)	Umitaku Maru (JAP)	Kaiyo Maru (JAP)
Kyoma Maru (JAP)		S.A. Africana II (SA)
Umitaku Maru (JAP)		Xing Yang Hong No. 10
Syonen (JAP)		(CHINA)

BIOMASS SHIPS *In all, 10 countries participated fully in the three Fibex and Sibex information-gathering programmes. The findings from these ships (left), were correlated at Cambridge, United Kingdom.*

OCEAN RESEARCHER *The British Antarctic survey ship, RRS* John Biscoe, *(below) was involved in the Sibex II campaign and continues its oceanographic research today, in addition to supplying cargo to BAS stations.*

MINERALS

It is generally agreed that the most significant issue confronting the Antarctic Treaty System is that of mineral resource exploitation. Reports from early explorers provided the first evidence that minerals might exist there but the issue was side-stepped when the Antarctic Treaty was negotiated, because of political difficulties and a lack of interest.

Decades of geological study have confirmed that Antarctica was once the central part of the Gondwana supercontinent and this has formed the basis for modern speculation about the continent's mineral wealth, extrapolating from known reserves on the continents that once lay adjacent to it.

On-shore mineral sites

Substantial deposits of iron in the Prince Charles Mountains and possibly the world's largest reserve of coal in the Transantarctic Mountains have been confirmed. Traces of numerous other minerals – gold, titanium, tin, copper, cobalt and uranium amongst them–have also been found in the Antarctic Peninsula and in other areas of the continent. Exploitable quantities of minerals have not been found, however, although knowledge of the Antarctic's mineral potential is sparse. Of the approximate 2

MINERAL-BEARING ROCKS *These geological specimens show traces of minerals present in the rocks. Although of scientific interest, they do not indicate the presence of commercially exploitable deposits.*

per cent of the continent not covered by ice, to date only 10 per cent has been mapped in detail and less than one per cent has been explored so far for mineral deposits.

On land the mineral site with the greatest potential is the Dufek Massif in the Pensacola Mountains, which extends over an area of 50,000km^2 (19,500mls^2). It is a geological formation known as a layered intrusion, formed when magma collects in a very large underground chamber and cools slowly, so that the minerals separate into layers as it solidifies. Some of the layered intrusions found on other continents, such as the Bushveld complex in South Africa, have proved to be among the richest mineral sites in the world.

Off-shore deposits

In recent years, interest has centred around potential off-shore oil and gas deposits, following the discovery by the US scientific drilling ship, *Glomar Challenger*, of gaseous hydrocarbons – which often occur with petroleum – in three out of four holes drilled in the Ross Sea continental shelf in 1973. This discovery more or less coincided

with the sudden rise in price of Middle Eastern oil. Hasty projections by the US Geological Survey, based on this speculative discovery, estimated that there were deposits amounting to 45,000 million barrels of oil and 3,250,000 million m^3 (115,000,000 million ft^3) of natural gas, although it considered only a third of this quantity recoverable. The USGS has since, however, disavowed this estimate, which it has acknowledged to have no scientific basis.

Owing to the harsh conditions, the thick ice cap and the environmental sensitivity of the region, exploitation of on-shore minerals is not seriously considered at the moment. The situation is different for off-shore hydrocarbons; existing technology could be used for exploration in such areas as the Ross Sea. Exploitation may require only a few advances in technology.

Vociferous debate

In the near future, any mineral exploitation would be uneconomic but, for some countries, strategic and political motivations might outweigh this consideration. At present, major companies, Antarctic

Quartz vein with golden-coloured pyrite (iron sulphide) crystals

Volcanic rock with brown crystals of sphalerite (zinc sulphide)

Volcanic rock and quartz vein with pyrite crystals and silvery-grey galena (lead sulphide)

Treaty members and environmental groups are locked into an increasingly vociferous debate about whether such exploitation should ever be allowed and, if so, under what conditions.

All the Antarctic Treaty powers believed that a regime would be easier to conclude before any exploitable quantities of minerals were discovered, and at the Ninth Consultative Meeting in London in 1977 they adopted a policy of "voluntary restraint".

They agreed to "urge their nationals and other states to refrain from all exploration and exploitation of Antarctic mineral resources while making progress towards the timely adoption of an agreed regime concerning Antarctic mineral resource activities".

Most authorities agree that while the extraction of the minerals on the continent is unlikely before the next century, off-shore surveying and even exploratory drilling for hydrocarbons might begin earlier than that; arguably it has already begun.

Impact of exploitation

The hazardous nature of the stormy Antarctic waters, with its sea-ice and icebergs, raises the possibility of oil pollution either from off-shore wells or, more likely, from tankers. Such leaks, which would be hard to plug in the extremely harsh conditions, would have a devastating and long-lasting impact on all levels of the fragile ecosystem, particularly on the ice-free coastal areas, where many birds, seals, fish and an abundance of smaller marine life are, of course, concentrated.

Off-shore oil exploitation would also require substantial on-shore facilities,

thus creating direct competition with wildlife for the limited amount of easily accessible, snow-free areas available and producing additional environmental disturbance to the region.

The effect of seismic surveying on oceanic wildlife is another area for concern. The high-powered shock waves used in such surveys may disrupt whale feeding and reproductive behaviour, and may also affect the behaviour of seals, fish and other marine organisms.

WIDESPREAD DEPOSITS *Although the extent of Antarctic mineral deposits is not fully known, numerous types have been discovered (left). In this view of the Beacon Supergroup in the Theron Mountains (below), clearly visible strata are exposed, some containing coal and shales.*

KEY

Ag	Silver	**Ni**	Nickel
Au	Gold	**Pb**	Lead
Co	Cobalt	**Pt**	Platinum
Cr	Chromium	**Sn**	Tin
Cu	Copper	**Ti**	Titanium
Fe	Iron	**U**	Uranium
Mn	Manganese	**Zn**	Zinc
Mo	Molybdenum		

Vein of white quartz and black magnetite (magnetic iron oxide) in volcanic rock

Bronze-coloured cubes of pyrite in greenish volcanic rock

Vein of black stibnite (antimony sulphide) and white barytes (barium sulphate) in granite

151

THE U.N. DEBATE

The idea that Antarctica should be placed under the trusteeship of the UN was a popular one in the immediate post-war period. (The Women's International League for Peace and Freedom recommended UN control over Antarctica as early as 1947.)

The case for placing Antarctica in the hands of the United Nations was first pressed by such people as Julian Huxley, the first director-general of UNESCO, who urged that it should organize an International Antarctic Research Institute. Lord Shackleton expressed the opinion that "there is no problem in the world which the United Nations is better suited to handle".

India pressed the issue several times in the late 1950s but by 1958 these proposals were overtaken by the negotiations for the Antarctic Treaty, which effectively precluded the idea of United Nations involvement.

UN agencies blocked
During the early 1970s, a number of UN agencies attempted to take up the issue but all attempts were blocked by the Treaty powers.

This did not stop a growing interest in Antarctica's resource potential among the developing nations. At the conclusion of the Law of the Sea negotiations in 1982, and in the light of the ongoing debates about an Antarctic minerals regime, further pressure was brought to bear by these countries.

Key speech
Their attitude was summed up in a key speech delivered to the UN General Assembly in October 1975 by Hamilton Shirley Amerasinghe, Sri Lanka's ambassador to the UN and the president of the Law of the Sea conference.

He said: "There are still areas of the planet where opportunities remain for constructive and peaceful cooperation on the part of the international community for the common good of all rather than the benefit of a few. Such an area is the Antarctic continent . . . where the now widely accepted ideas and concepts relating to international economic cooperation, with their special stress on the principle of equitable sharing of the world's resources, can find ample scope for application . . ."

First UN involvement
It was Malaysia who subsequently took the lead in raising the question of Antarctica at the UN during the 1982 opening session of the General Assembly. In mid-1983 they co-sponsored an initiative with Antigua-Barbuda, proposing that Antarctica be placed on the agenda of the 38th session of the General Assembly later that year.

At that Assembly, the UN's first real involvement in the Antarctic issue resulted in a consensus resolution requesting that the Secretary-General conduct a study covering all aspects of Antarctica and report back at the 39th session in 1984.

Crucial issues
Amongst the key issues raised by non-Treaty members were: broadening the decision-making base to achieve representation truly reflective of the international community; secrecy; that the resources of Antarctica should be declared the common heritage of mankind as is the case with resources of the seabed (beyond national jurisdiction); and outer space.

In September 1983, the Treaty nations responded to the growing power of this critical campaign by admitting Brazil and India to the ranks of the Consultative Parties, despite the fact that their scientific programmes were in their infancy and had not yet had time to produce results. In addition, acceding states were given observer status at a Consultative Meeting for the first time.

In 1984, the UN Secretariat duly produced a comprehensive report on all aspects of the Antarctic issue, which was debated over three days, but it added little to what had already been expressed. A consensus resolution was passed, expressing appreciation for the study and placing Antarctica on the agenda for the 1985 Assembly.

Divided Assembly
The Consultative Parties' position at the opening of the 1985 debate was that there should be no further debate on Antarctica in the UN; however, they promised in future to provide better information to UN agencies.

Malaysia countered with three resolutions: that the Secretary-General's study should be continued; that any mineral exploitation should be internationally managed, with an equitable sharing of the benefits; that South Africa should be excluded from participation in Consultative Meetings.

The vote on these points was overwhelmingly in favour but the majority of Consultative and Non-Consultative Parties refused to participate.

At the 1987 debate, the divisions between the ATCPs and the Non-Aligned Movement continued. Voting was again: in favour of excluding South Africa; for a minerals moritorium; and for the Secretary-General to be invited to all Treaty meetings.

Future working relationship?
Whatever the outcome of these UN debates over the next decade, they have provided the first real point of contact between the Antarctic Treaty powers and the international community at large. Hopefully, these meetings will provide the foundation for a productive relationship in the future.

A SCIENTIST'S VIEW

Born in 1926, Charles Swithinbank became fascinated with the polar regions at the age of eight, when his mother read him stories of exploration. Now, more than 50 years later, he is still making regular visits to the continent that so attracted him.

Having already visited the Arctic as a sub-lieutenant in the Royal Naval Volunteer Reserve (1944-46), Charles Swithinbank went to study at Oxford University (1946-49). There he became secretary of the university Exploration Club and took part in expeditions to Iceland and the Gambia River in West Africa during the long summer vacations. His few weeks on the ice cap in Iceland resulted in the publication of two articles in scientific journals. He says: "that gave me the confidence to realize that you don't have to be a genius to contribute to knowledge".

When the Norwegian-British-Swedish Antarctic expedition (1949-52) was being organized they wanted four people from the UK. As chance would have it, he was offered the opportunity to join and was signed on as assistant glaciologist. He was to spend two-and-a-half years without a break at a place called Maudheim in Queen Maud Land and gained his PhD in polar glaciology as a result.

His first experience of Antarctica left him with many strong impressions.

"The thrill of treading new ground – which in the interior of the Antarctic is still, even today, relatively commonplace – was great. The idea that everything you were seeing was new to man and therefore had to be recorded meant that you were extremely hard at work writing notes and observations over a wider field than you would be nowadays.

"People have got more and more specialized now, whereas at that time we only had a few people and we didn't have all kinds of specialists. We didn't have a botanist or an entomologist on that expedition but we made jolly sure we collected all the mosses, lichens and mites.

GLACIOLOGIST, CHARLES SWITHINBANK

"I was there as a glaciologist but a glaciologist needs a map to plot his observations and there was no map, so we were making maps at the same time. I remember standing on a mountain peak surveying with a theodolite. The doctor on the expedition, a Swede who was acting as my field assistant, was writing notes while I dictated the numbers. Getting rather bored, he turned a rock over and there were little red mites running about underneath, a variety that was new to science and it's now got his name attached to it: *Maudheimia wilsoni*. Chance discoveries like that were happening the whole time.

"During my first season there, I remember dog sledging in April (Antarctic autumn) with a Norwegian in −40°C (−40°F). I had never been in such cold weather in my life. Harnessing up dogs at those temperatures is distinctly uncomfortable because your fingers keep on getting frozen.

"We ran low on food because, in dog sledging, your distance is inversely proportional to the weight you're carrying. You will go faster and further if you're carrying less weight. On the other hand, you don't want to run out of food and you have got to have a very fine balance between the two. One of the safety factors in travelling with dogs is that you have far more dog food than man food. In an emergency, you turn to eating dog food first, in our case, dried Norwegian cod fish. Your ultimate sanction is to eat the dogs, which we never did. We weren't that badly organized or we weren't cold blooded in the way that Amundsen was in feeding dogs to dogs."

The expedition, dedicated as it was to science rather than to pure exploration, was widely acknowledged later as being a model for the IGY.

After a summer in Greenland and a year at the University of Stockholm, he spent four years studying the distribution of pack-ice in the Northwest Passage for the Canadian Government, a job he took just before being offered the opportunity to join Sir Vivian Fuchs' Commonwealth Trans-Antarctic Expedition.

For three seasons in a row, from 1959 to 1962, he was leader of the US Antarctic Research Program on the Ross Ice Shelf and, from 1963 to 1965, he was the first British exchange scientist on the Soviet Antarctic Expedition. He is thus well qualified to talk about the difference between attitudes to science.

"In the case of the Russians, there was a smaller proportion of individual scientists, with a proposal to do good science which had been approved by the funding bodies, than there were doing standardized observatory sciences.

"The American programme had people signed on to do standardized observations but a greater proportion of Americans were doing new projects as a result of their own initiative, which involved less tight control of exactly what they were measuring or observing."

The trend of the last 30 years has been towards increasing scientific specialization, a trend which he says is necessary but occasionally saddens him. Whilst primarily a glaciologist, his early experience taught him always to keep his eyes open for discoveries in other areas of science.

He recalls: "In 1975 I was cruising along in a Twin Otter plane with five of us on board, all intent on radar-sounding of the thickness of the ice, and we came across a nunatak sticking through in an area where there was nothing on the map. It was, in fact, a very long distance in any direction to any other rock. So I immediately said, let's land and take samples; those rocks turned out to be the oldest rocks ever found in the Antarctic Peninsula, a thousand million years old, in fact. If I had ignored that and just said, well I'm a glaciologist and we're here to measure ice thickness, those rocks might not have been sampled to this day and so an important gap in geological knowledge would not have been filled."

Swithinbank Glacier, Moraine, Range and Slope; Charles Glacier and Nunataks – he has certainly left his mark on Antarctica.

"I had nothing to do with it. These things are always done behind your back, and if anyone believed that you were trying to get something named after yourself there would be an automatic disqualification.

The only thing I've ever done is to be in the right place at the right time. I have six place-names altogether, two named after me by the New Zealanders and later recognized by everyone else, three named after me by the Norwegians who were responsible for the places named on the Norwegian-British-Swedish expedition, and one by BAS.

"The glacier was named after me because I was the first person on it and I measured its rate of movement when I was assistant glaciologist on the Norwegian-British-Swedish expedition.

"The range and the moraine were named by Wally Herbert, the chap who later drove dogs from Alaska to Spitsbergen across the North Pole. He was making the first maps of the area, working for the New Zealanders in the season in which I was working for the Americans. We were the only people in that part of the Transantarctic Mountains and he was searching around for names to put on features he'd discovered."

The whole question of naming places can be a saga in itself.

"In the case of the nunatak I mentioned, it was given an American name – Haag Nunatak – because Finn Ronne had seen it from a great distance and mapped it in a very wrong position. Later, somebody had been to the map position and found nothing and so it had been wiped off the map. When we found it, there was no reason to believe that anybody had ever seen it before. There have been a lot of cases like that – of people giving place-names to things seen by earlier explorers.

"The place-name authorities now are pretty strict about asking how precisely you have fixed the position of this rock and to what extent you can guarantee that nobody has already seen it and put it on an inaccurate map. That is the skill of place-naming. Immense confusion will arise later if you pepper place-names on the landscape without maps of sufficient accuracy to describe unambiguously what has been discovered.

"There can be no central authority for this because of the dispute over sovereignty. There are national place-name authorities in probably every country working in the Antarctic but, because

of the situation of claimants and non-claimants and those who do not recognize claims, it is something to which you could not get agreement internationally.

"So, sensibly, the Treaty has avoided the subject because nationalism and national prestige are involved; it's a very tricky subject to get agreement on. All the claimants whose claims do not overlap recognize each other's place-name authorities but, in the Antarctic Peninsula, where you have Argentine, Chilean and British claims overlapping, there is no way that you can resolve the problem.

"In spite of the sovereignty difficulties, all place-naming authorities try very hard to perpetuate the names given by the first explorers if they know about them and if they can recognize what was named.

"The British and Americans both have permanent Antarctic place-name committees and they now agree, before putting on maps, 99 per cent of all place-names. The 1 per cent is in cases not so much of disputed first sighting as in appropriateness. What are you describing? Do you call it a peak, a promontory, a peninsula, an ice rise?

"When the first sighting is disputed, nations tend to favour using their own name."

"The principal disagreement that caused the British and Americans to go their own way – it sounded silly to the rest of the world – was resolved 25 years ago. The Americans called the Antarctic Peninsula Palmer Peninsula; the British called it Graham Land. The argument dragged on because it revolved around the question of who first sighted Antarctica.

"That last point is still in dispute because of the extraordinary coincidence that British, American and Russian explorers may well have sighted it in the same year – early 1820. Since they didn't record properly what they saw, it's anybody's guess who was the first of those. None of them said 'we have discovered Antarctica' because they didn't even know there was a continent there to discover. So the problem comes with later generations trying to use hindsight on something which doesn't bear hindsight because no explorer at the time knew the significance of what he saw.

"Anyway, Palmer was the American candidate and Graham was the British. The lack of agreement was silly. Here we both were, operating in the same area, describing it in different ways.

"I think it was Paul Siple – the boy scout with Byrd's first expedition in 1929, who was an old and respected scientist in the 1960s – who said why not lets split this Peninsula and call the northern half of it Graham Land and the southern half Palmer Land. That went through the committees without difficulty and we have stuck with it to this day. So the biggest sticking point between the American and British communities was resolved in one fell swoop.

"But what the British and Americans now call Graham Land, the Chileans call O'Higgins Land because he was their national hero and it's Argentina Antarctica to the Argentines. There's going to be no resolution between those parties as far as I can see. So the Treaty powers have very sensibly accepted that it is a sensitive subject. Arguing on such matters could destroy the collaboration about things that are more important."

The last time Charles Swithinbank visited the continent was in November 1987, the year after he retired, having held the post of Head of Earth Sciences at BAS since 1974, where he had 50 scientists working for him.

What human qualities does Charles Swithinbank consider necessary to work and live successfully in Antarctica?

"Big subject. The ability to work with other people is very important in that you can never go it alone. You sometimes see a conflict between the ambitious scientist and the fact that he has to collaborate with people who are equally ambitious in order to get his ambitions fulfilled. That can be a situation of potential conflict but anyone intelligent will realize that; since conflict kills collaboration, you had better avoid it as far as you possibly can.

"You need the ability to live and work with people even under stress. The stress is not enormous but is made enormous by people insisting on getting a lot done in a short time. So, if you are out in bad weather and the wind is shrieking, you're automatically more impatient of your colleagues' failings because you are already annoyed at the weather.

"What I have noticed in living with people is that there's no preconceived type that does better than any other. I've seen total loners and people who are the life and soul of the party do well as long as they pull their weight, particularly at things they don't want to do but have to do, such as sweeping the floor. You very quickly detect shirkers.

"It's really not worthwhile gaining the disrespect of your colleagues because it will cost you dear when you need help. It's interesting. Very few people in the Antarctic proffer help but you know that if you ask for help and it's perceived that you need it, you'll get it.

I observed very early on, starting on an international expedition, that if there's anything which causes friction it's personality and not nationality. I cannot identify any nationality I would prefer to be with than any other.

"Working in Antarctica makes you conscious of the tremendous privilege that you've had. Last year in the Antarctic I met some mountaineers who had paid almost $20,000 for 10 days in the Antarctic and they were seeing a tiny, tiny fraction of what I've seen.

"A degree of humility also comes from seeing a mountain range that nobody has worked in before, that has been there for 1,000 million years, and saying to yourself I am in here for a moment of time in terms of this mountain.

"People who have had accidents in Antarctica have often failed to recognize the dangers."

"On our first expedition we lost three people. Every single accident I've ever heard of has been the fault of the people concerned. In that case, there was one person driving a vehicle too fast in a fog and he drove over the ice front into the sea. There were four people on board; one survived by swimming to an ice floe, the other three drowned.

"There are inherent risks. You cross thousands of crevasses without stopping to probe them until they get too close to each other. Then you stop and probe and accept the delay. What kills people is the impatience in going on travelling when there's overcast conditions, a whiteout, because then you can't see a snow bridge. All of a sudden the ground may disappear from under you. Had you waited for the sunshine, which can be extremely frustrating, you would have survived.

"So the risks are very much in your own control. There's this conflict all the time: reduce risks to almost zero and you get almost zero done; increase them too high and you won't bring back the results."

LONE SILHOUETTE *The Twin Otter wings its way across the Antarctic Peninsula, casting a shadow on the snow fields below.*

GREENPEACE PERSPECTIVES

"*Yesterday we were able to get out of the base for the first time in a number of days. The sight we beheld was one of indescribable beauty. The glow of sunlight refracted around the curve of the Earth's atmosphere, back-lighting the Barne Glacier. There were reds and crimsons in the far distance, and dark silhouettes of distant lenticular clouds, all under a dome of deep blue space, speckled with stars.*"

Justin Farrely, Greenpeace base member

WORLD PARK STRATEGY

Greenpeace and other members of the Antarctic and Southern Ocean Coalition (ASOC) use the term "World Park" to describe a status which would for ever protect the natural environment of Antarctica from human depredation.

The first serious suggestion of a World Park regime for Antarctica was made at the Second World Conference on National Parks, held in Yellowstone Park in 1972.

It agreed to a resolution acknowledging "the great scientific and aesthetic value of the unaltered natural ecosystems of the Antarctic" and called upon all of the ATCPs to "negotiate to establish the Antarctic Continent and the surrounding seas as the first world park, under the auspices of the UN".

Three years later, New Zealand announced its support for the World Park concept and gained some backing from Chile, though the backing was conditional on Chile not having to give up its territorial claims under such a regime. No other ATCP supported New Zealand's lead and the question has never been debated or considered at a Consultative Meeting.

The present attitude of the Antarctic Treaty states is that the question of World Park status for Antarctica is not even worth considering.

Implementing the principles

The World Park concept is based on the following set of principles:

1 The wilderness values of Antarctica should be protected;

2 There should be complete protection for Antarctic wildlife (though limited fishing would be permissible);

3 Antarctica should remain a zone of limited scientific activity, with cooperation and coordination between scientists of all nations;

4 Antarctica should remain a zone of peace, free of all weapons.

Greenpeace believes there is no intrinsic need to establish a new legal entity to implement these four principles. The necessary decisions could be taken through the existing framework established by the Antarctic Treaty System, CCAMLR, the Convention for the Conservation of Antarctic Seals and the Agreed Measures for the Conservation of Antarctic Flora and Fauna.

YOUTH IN ACTION
A young campaigner announces his support for the World Park scheme.

If it were felt necessary to extend some of the existing environmental measures to cover the more rigid requirements of a World Park, this could be done by the negotiation of an Antarctic Conservation Convention as part of the Antarctic Treaty System.

The World Heritage Convention

In addition to utilizing the Antarctic Treaty, there is another viable international mechanism for the establishment of a better protection status for Antarctica, which could be a first step towards meeting Greenpeace's World Park criteria.

In 1972, 60 nations participated in the drafting of the Convention Concerning the Protection of the World Cultural and Natural Heritage (World Heritage Convention). Since then, more that 90 nations have ratified the convention and over 160 World Heritage Sites have been named.

A major objective of the convention is the promotion of global awareness, and the protection of natural and cultural sites of "outstanding universal value".

For inclusion on the list, a site must fulfil at least one of the following criteria: represent the major stages of the Earth's evolutionary history; represent significant ongoing geological processes, biological evolution and humans' interaction with the natural environment; contain superlative natural phenomena, formations or features or areas of exceptional natural beauty; or contain the most important and significant natural habitats where threatened species of animals or plants of outstanding universal value still survive. Antarctica fulfils all four.

The World Heritage approach would permit all nations to work constructively with the nations actively working in

MAKE ANTARTICA A WORLD PARK

A VITAL STEP FORWARD *Giving Antarctica and its icy seas the status of a World Heritage Site would afford at least a measure of protection to the continent's fragile environment.*

Antarctica for the protection of the continent. Ideally, the Treaty nations themselves should nominate the entire Antarctic continent for protection status under the convention.

The effects of the scheme

A World Park management scheme represents only a modest departure from current practices in some cases. In others, the differences are decidedly more dramatic.

The area would remain demilitarized and free of nuclear activities. Scientific research would be given top priority, enabling scientists to develop a long-term plan for utilizing the continent's unique attributes for the benefit of the entire global community. There would be greater coordination of scientific programmes than at present, including a fuller sharing of existing base facilities amongst a wider group of nations.

Mineral exploration and exploitation would not be permitted. Geophysical research at the boundaries of pure science and prospecting would be allowed but all data would have to be in the public domain.

The administration of Antarctica's marine resources would be undertaken according to the "ecosystem approach" established under CCAMLR. Some changes, however, might be necessary to ensure that CCAMLR's ecosystem management concept worked as the convention intended.

Tourism would continue under carefully monitored and controlled conditions but "colonization" by persons who did not have specific or justifiable tasks would be prohibited. Sovereignty would not be affected.

Enforcing the rules

The Antarctic Treaty System relies on cooperation by member states to uphold its rules. This has not worked to the advantage of the Antarctic environment. However, if the World Park concept is implemented, it will be meaningful and effective only if an enforcement mechanism is created.

The first step towards enforcement would be the creation of an Antarctic Environmental Protection Agency (AEPA) which would: undertake independent investigations and assessments of proposed activities, both scientific and logistic; conduct inspections and monitor operations; and prepare environmental regulations for all activities taking place on the continent.

Since 1970, over 110 nations have set up high-level environmental protection agencies. They have been needed to balance other, more powerful, interests and to ensure that the long-term health of the environment is properly weighed in decisions concerning development.

If the rules of an Antarctic World Park were violated, the matter could be turned over to an Infractions Committee, which would assess the evidence and recommend appropriate action.

VOYAGE TO ANTARCTICA

In January 1987 the MV Greenpeace *set sail for Antarctica to establish the first Greenpeace base. Campaign Assistant, Maj de Poorter was on board. Her account of the voyage is reproduced here, followed by extracts from the diaries written during the course of the year by the four base members.*

21 NOVEMBER 1986; AUCKLAND, NEW ZEALAND

Arrived at Auckland International Airport. It is difficult to believe that in as little as six weeks' time I will be on the MV *Greenpeace*, the 55m (180ft) expedition ship, sailing towards Antarctica.

The aim for this season: to build the first non-governmental, permanent scientific base in the Antarctic at Cape Evans. Four Greenpeace volunteers – three men and one woman – will then spend the harsh winter at the base, where they will carry out marine biology research and investigate the human impact on the Antarctic. The ultimate aim: to urge forward the declaration of Antarctica as a World Park, where the use of natural resources to commercial ends will be prohibited. Seems logical? For most people, yes, but not for the heads of the

AN HISTORIC VOYAGE
The MV Greenpeace *leaves Lyttelton, New Zealand (below). Among the crew was Maj de Poorter (left) who wrote the account of the voyage that appears here.*

Treaty states, who "manage" Antarctica. They are at this moment busily negotiating the mineral exploitation of Antarctica – behind closed doors, of course. We must put a stop to this before we lose the last paradise on Earth.

15 JANUARY 1987; ROSS SEA, ANTARCTICA

The first few days at sea have been rather rough, at least for land-rats like myself. The Roaring Forties and Furious Fifties certainly lived up to their names. It felt like a giant roller-coaster behind the steering wheel at times – "hang on" was the message.

Albatrosses and mollymawks – master sailors – regularly followed us to begin with. Further south, the

species changed. After a few days, the albatrosses were replaced by prions—giant petrels and Cape pigeons – a sure sign that we were crossing the Antarctic Convergence, the border between the warm northern and cold southern waters. Pretty soon after that we saw the first iceberg, and on 12 January we crossed the Antarctic Circle. We are living in continuous daylight now.

In the Ross Sea there is a ring of ice in the north and a second band of dense pack-ice near the Antarctic continent in the south, with open water in between. Watching out for icebergs, bergy bits, and especially for growlers – the worn-down icebergs that are almost completely hidden below the surface of the sea – we entered the northern ring of ice. A few hours later we were through it and in open water.

Immediately we saw our first penguins. Like dolphins, they "jump" in and out of the water in a black-and-white flash. Like us, they are heading south, although at a much greater speed.

19 JANUARY 1987: LEWIS BAY

Today we tried to make our way through the remaining pack-ice between us and Cape Evans. The MV *Greenpeace* slowly made its way through 2/10th-ice thickness (ice covered 2/10th of the surface), towards open water only 1.5km (1ml) or so away, which would allow us to round Cape Byrd and get close to Cape Evans.

Everything went well to begin with but after a while the ice started to close in more and Jim, the captain, decided to play extremely safe and evacuate five people from the ship by helicopter. It was more a testing of the emergency procedures than an evacuation, but, to be honest, it was a bit scary none the less. At dinner time everyone was back on board, and the MV *Greenpeace* was back in open water. Unfortunately we were still in Lewis Bay, though, the "wrong" side of Ross Island.

24 JANUARY 1987: LEWIS BAY, ROSS ISLAND

We have been in Lewis Bay, at the north side of Ross Island, for the last nine days. Mount Terror and Mount Erebus tower

INTO THE ICE *The MV* Greenpeace *enters the Ross Sea pack-ice (above). Later, the first supplies are transferred from the ship to the base site by helicopter (below).*

before us a few miles away. Cape Evans is still about 70km (45mls) away on the other side of Ross Island. A few people have already been dropped there by helicopter to start the preparatory work – measuring out and levelling the building site, rigging up the radio mast. They cannot do much more until the ship is closer to Cape Evans but the pack-ice is much too dense at the moment.

It is so breathtakingly beautiful here. A pod of orcas – 10 or so – passes close by this ship regularly. First you see the small black triangles – the tops of the dorsal fins; then, slowly, the rest of their fins emerge, followed by their glistening backs; then, all too soon, you hear "pffs" of breath before they dive under again graciously. We passed an ice floe containing a cluster of Adélie penguins and a lazy leopard seal, which did not even open its eyes as we steamed by. As I write, the water laps comfortably against the bow while, somewhere in the distance, a penguin lets out a melancholy cry.

25 JANUARY 1987: LEWIS BAY, CAPE EVANS

Yesterday at 9 p.m., the helicopter managed to take off with Jim, our skipper, on board to take a look at the ice situation. He came back with good news. "There is a lead through the pack around Cape Bird, then a short stretch of 3/10th-thickness ice, and behind that nothing but open water to Cape Evans," he exclaimed, triumphantly. The conclusion: "Let's go!"

The stretch through the actual pack was thrilling. A school of orcas swam gracefully towards us as we entered the narrow lead. Everyone stayed on deck, all thoughts of sleeping forgotten! Then, at a quarter past midnight, the ship resounded with the loud wail of its siren. We were out of the pack-ice with open water between us and our destination. The reaction on board was tremendous. Everyone was smiling from ear to ear and hugging everyone else and there was an immense feeling of relief and unity. None of us doubted that we would succeed in all our objectives now.

At a quarter past eleven this morning we anchored in 16 fathoms of water, off the Cape Evans beach. We had arrived and now we could get on with the work. Just after lunch the first sling-load was transported by helicopter.

30 JANUARY 1987: CAPE EVANS

The floor and walls of the base are up and standing. Garry, one of the helicopter pilots, did a marvellous job today. He airlifted each of the 22 roof panels into place, exactly where they should be – a real precision job. A funny thing, though: when the roof was finished, a ladder was discovered in Cornelius' room. It had been lowered into place by helicopter early on and then forgotten. Now the roof is in position and no amount of pushing will persuade it to pass through the doorway!

The base is taking shape.

2 FEBRUARY 1987: CAPE EVANS

I spent most of today on shore at Cape Evans with Gudrun, the German biologist, who will over-winter here for Greenpeace. In the middle of West Beach (about 400m / 1,300ft away from Home Beach, where the base is built) there is a pile of bones and a mummified seal skin. In the middle of this "fortress" is a skua chick. The parents started diving at us, but we kept a good distance from the nest and they soon left us alone. On the beach we saw some red algae and some sponges; a bit further along we came across a single Adélie penguin, basking in the sun on top of the rocks.

On the way back we passed Scott's hut, which is at the south end of Home Beach. Built in 1911, it was here that Robert Falcon Scott and his companions wintered before starting on their long trek to the Pole. A strange atmosphere hangs around the hut. The rough wooden skis still rest against the wall, the mummified body of a sledge-dog sits at the door, and bales of hay for the ponies are piled high nearby. I wondered what it was like inside. Although it is no problem for visitors from McMurdo or for tourist groups to obtain the key to the hut from New Zealand Scott base's Officer-in-Charge, it becomes an impossibility if members of Greenpeace ask for it. McMurdo shares the same attitude: officials there refuse to pass on local weather information if our helicopter is in the air, even if it is on its way to McMurdo.

STANDING BY *The MV* Greenpeace *lies anchored off the Cape Evans beach close to the Greenpeace base.*

9 FEBRUARY 1987; CAPE EVANS, SCOTT AND McMURDO BASES

Today saw the start of "World Park Antarctica Week". Greenpeace took the opportunity to visit the New Zealand Scott base and the American McMurdo base, about 20km (12mls) away from Cape Evans. We wanted to see how the bases treat their rubbish and effluents.

The US base, McMurdo, has been here for more than 25 years, yet it still looks like a giant messy construction site. We saw a gaping hole in the mountain flank where the nuclear power plant used to be (it was dismantled because there were too many technical defects), and a giant rubbish dump, containing plastic and rubber items, steel pipes, a battery, and trucks – it looked as though everything was thrown out, even if it could still be used or repaired. All junk should be removed from the Antarctic. If you cannot find a way to take your rubbish home with you, you shouldn't be here in the first place. A bit further down the road, near the ice-wharf, we found a similar scene: a pile of truck skeletons, wheels, oil drums, and a pipe discharging brightly coloured liquid straight into the water.

A lot will have to change here. It is shameful and totally unjustifiable to create such a mess when you are surrounded by such a beautiful environment.

NO REST FOR THE WEARY *Greenpeace members work long hours to erect the base at Cape Evans (above), fully aware of the need for urgency. Setting up a radio mast (right) is a priority.*

11 FEBRUARY 1987; CAPE EVANS

Wind: 17-22m per sec (35-50mph). Temperature: −15°C to −11°C (5°F to 12°F). Snow sweeps over the deck. At the base camp they are only able to work inside, and we all stay below deck.

From the information we have received about the ice conditions, it appears that the circle of pack-ice between ourselves and the open water to the north is thickening again. We must leave soon.

13 FEBRUARY 1987; CAPE EVANS

The construction of the base is finished, and today we held the house-warming.

Inside the base there are four tiny bedrooms. Justin has got all the radio, telex and other communications equipment in his, and Gudrun has all her scientific equipment with her – the two rooms are almost filled up already. Then there is a communal space with a kitchen and a lounge. In the lounge are a few cosy chairs, a TV/video, and a good stereo installation. Alongside is a storage room, and a small but comfortable bathroom with shower. The toilet system generates a small volume of waste, which is bacteriologically inactive, so everything can be stored and taken away by ship. (We intend to store all our garbage and return it to New Zealand next year.)

14 FEBRUARY 1987; CAPE EVANS

The four winterers spent their last afternoon on board. After dinner, they said their farewells before returning to the Greenpeace base. It is difficult to imagine that these four people, who have grown to be such good friends over the last few months, will now really stay here in the Antarctic for a whole year. Everyone hugged them and wished them good luck.

Once they were ashore, the MV *Greenpeace* set off towards Cape Byrd. The four stood on the roof of their "home", waving to us, and on the ship everyone was on deck, waving back and choking back the tears. *Partir, c'est mourir un peu.*

163

The four over-winterers – Kevin Congalen, Gudrun Gaudian, Justin Farrely and Cornelius van Dorp – took it in turns to write the Greenpeace base diaries. Their long sojourn ended in January 1988 when four new Greenpeace volunteers took their places.

29 MARCH 1987:
Gudrun Gaudian

Today was such a beautiful day. Mount Erebus was spitting plumes of smoke into an immaculate blue sky. Not a sound could be heard: the vast, white silence of Antarctica. The Transantarctic Mountains loomed, clear and close-seeming. They are over 60 nautical miles away but it seemed far less because of the incredible clarity of the air. All that could be heard was the occasional creaking of the ice floes. The sea has been freezing over for the past week or so and the ice is becoming quite thick, although not thick enough to prevent the tidal movements of the water beneath the floes breaking them into smaller pieces. It makes an eerie sound, just like an old door in a secret castle being opened very slowly. A Ross seal lolled about in a tide crack, his head poking through the slushy water.

Eventually the sun set, tinting the snow, the ice slopes, glaciers, and rocks the most amazing shade of red.

6 APRIL 1987:
Cornelius van Dorp

At the moment I'm lying on my bunk thanking my lucky stars, and our good leader (he prefers this title to "base leader") Kevin, that we're safely home in our little green box and not out in the blizzard. Because that's what it is out there – our first genuine "Herbie". The wind is howling from the south at up to 70 knots and for the first time large snow drifts are building up outside, burying our poor little Kobota tractor and everything else standing outside. The temperature out there is −26°C (−15°F), but in here

ALONE FOR THE WINTER
The base completed (right), the ship leaves the four volunteers for the winter. The team (below) comprises, from left to right: Cornelius van Dorp, base doctor; Justin Farrely, radio operator and technician; Gudrun Gaudian, scientist; and Kevin Conaglen, mechanic and team leader.

it's jeans and tee-shirt weather at 20°C (68°F) – thank God!

I wish I could convey to you the supreme comfort I feel just now with every muscle aching and the blisters on my feet smarting from our 72km (45mls), three-day marathon to Scott base and back. The plan was to stay and socialize for a few days but something made us push our aching bodies back to Cape Evans, despite the fine weather – cloudless skies and no wind. Sure enough, just a few hours after our return the black clouds began rolling in and now the wind is howling.

The walk to Scott base was our first tentative excursion, and was as much a test of our gear as a mail run. The trip was preceded by five days of brilliant clear weather, during which time the whole of McMurdo Sound, as far as we could see, had frozen over with about 10 to 15cm (4 to 6in) of ice.

We started out by skirting around the edge of the old ice, which had not broken out this year, but we soon saw that we would save quite a few miles by walking over the new ice. After testing the ice thoroughly with our axes we decided that it would support our weight and strapped on our skis. We had only gone several hundred metres, however, when I thought I heard something and motioned for Kevin to stop. We strained our ears. There it was again – a powerful "whoosh" of expelled air under the ice. Kevin's face went white as a few seconds later on our other side and much, much closer came the same noise. "Orcas," he shouted, "and they're hunting. Let's get out of here." We both turned and zig-zagged back towards the thicker ice as fast as our skis would take us. It took 10 or 15 minutes of strenuous effort to reach it, and we were tracked the whole way by the orcas. Their characteristic sounds 15cm (6in) under our feet lent wings to our skis. They may have been trying to be friendly, I don't know, but we both had the terrifying feeling that we were being hunted. From then on we stuck to the safety of the crumbled old ice, very soon discarding our skis and continuing on foot.

When we passed this spot on the way back several Ross seals had made breathing holes in the ice and one of them had a bleeding gash near its right eye. Contrary to the usual placid nature of these animals, this one reared up, snorting

SCIENTIFIC PROGRAMME *Gudrun marks ice samples taken outside the base. The data will be analysed by specialists.*

and blowing aggressively, as we approached. Nearby, to confirm our experience with the orcas, there were several areas of thin ice where it was evident that a massive force had pushed up and broken the surface from below.

The experience of living here is mind-expanding to say the least. The pristine beauty of the still, clear days is made all the more vivid by the preceding days of furious 50-knot winds. The vast silence you experience when you get away from the base is purifying to the soul.

13 APRIL 1987;
Kevin Congalen

Most days we have the same old routine jobs to do: warm the Kobota before starting it (the temperatures are around −22° to −28°C/−8° to −18°F, which makes the oil quite thick), top up the snow melter, refuel the engine room tanks (every three days) and the Rayburn tank (which only has to be done every 26 days or so), and empty the slush buckets and toilet tank.

Justin's leisure time seems to be spent reading science fiction novels and modifying things, from electronic equipment to skis. Gudrun reads mostly political and world affairs books, and does a lot of sketching and watercolour painting. Cornelius is right into deep thought and mind books, plus he spends time strumming on his guitar. My leisure time has gone into oil painting, pastels, and modifying our clothing and tents on the industrial sewing machine (we now

have rainbows on most of the field equipment). I like to read books on adventure and travel, too, and all of us really enjoy going for walks around the beach and the cape here.

I like to look at the crimson sky and the violet sea, and to feel the cold ice and strong wind on my face. It's all these elements that make this beautiful parlour of ice and snow. It is the black of a winter's night that makes you appreciate a summer's day. This is the only continent left on Earth that can remind people how clean the planet should be.

19 APRIL 1987;
Gudrun Gaudian

I'll start with the trip to Mount Erebus glacier ice tongue. We were going to start dead early in the morning of Tuesday, 7 April, but it took a couple of hours after waking before we were on our way across the hills towards the frozen sound. We skied across the ice for four hours.

The beauty of the scenery was something else. We saw sundogs, wonderfully lit mountains, expanses of ice like deserts, and clear open sky. We passed big Razorback Island, which loomed like an Egyptian mummy, and finally the ice tongue itself, with its ice caves and weird ice formations. That night we were cold and tired by the time we erected our tent.

The next five days were spent blissfully lolling about, checking out the area, dodging crevasses, and snuggling into our sleeping bags while a blizzard was howling away outside.

It was good to get back to the base, though. We arrived last Saturday afternoon after a three-hour slog across the ice. The promise of a long, hot shower spurred us on and we travelled back without stopping.

That was a week ago. Since then I have been working in my room most of the time, analysing samples, reading scientific papers, taking regular meteorological readings, writing, and observing the local wildlife (now reduced to a few Weddell seals – skuas and penguins have truly gone). The sea is freezing over well and I will soon be able to go out fishing on the ice, unless a storm pushes the sheet ice northwards.

Soon the sun will be gone; in fact, there are only another four days to go before she will set for a few months. I wonder what it will be like – perpetual darkness? No sunsets, no red tinges on the mountains and clouds, only the moonlight to illuminate my path when I go out sampling, inspecting the beaches for wildlife, and ice fishing. I am looking forward to it.

26 APRIL 1987: Justin Farrely

At the moment, Kevin and Cornelius are working to get both our skidoo vehicles ready for use on the pack-ice. The water in the main sound has still not frozen for good. The pack-ice is there when there is no wind, but as soon as the wind rises, away it drifts to stack up on the southern Victoria Land shores or blow out to sea. We may have to wait another two or three months before we can be sure that the ice will stay put, and it is safe to travel on it. Towards Scott base and McMurdo the situation is a lot more stable. The ice is over 20cm (8in) thick in most places, and locked in by several islands.

Gudrun and I have surveyed the ice out that way and I shall be helping her to cut and maintain several holes in the ice for her fish and krill population observations. She has promised me a fillet of fish for my efforts. The question I ask myself is what will she catch? I can see me trying to eat a krill fillet the size of my thumbnail. We have to lug all the machinery over to the ice sheet so that we can start work. First, we shall use an auger to drill through in several spots so that Gudrun can ascertain the water depth. When she has decided which are the most suitable spots to drill, we shall use a chainsaw with a 2m (6ft) blade to cut the actual holes, and a tripod jack to remove the ice plugs. The ice refreezes at a rate of about 2.5cm (1in) per day, so if we go out at least every two days we should have no problem maintaining an open patch of water. However, the weather may cause difficulties. If it blows consistently, particularly if Gudrun has to spend any length of time at an ice hole, we may have to build a structure over one or more of the holes to keep her warm.

We do not have enough lights to put up outside to guide us in the filling of the snow melter. Gudrun and I are working

23 MARCH 1987
ON THE ICE

JUNE 1987
OUTSIDE AFTER BLIZZARD

SWEET MEMORIES *This series of slides captures some of the memorable moments of the first Greenpeace year in Antarctica.*

on ways of locating the things we need, and placing them in sensible places. It is good at the moment, as we can do trial runs for a lot of our regular jobs at "night", and fix problems during the day, as there is still a reasonable amount of light about – if somewhat dim. The weather is worsening, and now we are into the $-26°C$ to $-29°C$ ($-14°$ to $-22°F$). In fact, today could almost have been a $-30°C$ ($-23°F$) if you looked at the thermometer at the right time. Whenever you travel outside, your breath freezes to the lapels of your jacket. Without her hood, Gudrun's hair freezes white at the ends, and her breath forms beautiful crystalline structures of ice.

We have now switched off our freezers, as there seems little point in leaving them on when they cannot get down to the $-20°C$s (below $-5°F$) anyway.

1 MAY 1987: Cornelius van Dorp

The sun disappeared about two weeks ago, although there is still a definite daytime with a few hours of twilight between 1000 and 1500 hours. The last sunset was quite wonderful, even more so as it was our first glimpse of the sun for 10

days. I can still see the great golden orb skimming along the top of the Barne Glacier, sinking out of sight, then emerging past the end of it to sink into the sea. The next day was clear but the sun did not quite make it over the horizon.

In clear weather the sky is a brilliant canopy of stars, some of them friendly and familiar. The Southern Cross is almost directly overhead with Canopus and Sirius, and sometimes, low on the horizon, you can catch a glimpse of Orion. I've just come back from a walk outside – a brilliant star-lit scene, not a breath of wind, the temperature at $-30°C$ ($-23°F$) and in the sky for the first time the ghostly transparent streamers of the beautiful aurora australis.

We have created a very comfortable yet practical home and we now are finding our niches.

10 MAY 1987: Kevin Congalen

This last week has been rather quiet. Justin and Gudrun have been busy trying 101 ways to cut a 1m- (3ft-) round hole in the sea-ice in order that Gudrun can set some of her fishing lines and nets to start taking samples and analysing them. The rest of us are hoping that there will be some samples left over from her analysis, so that we can have fresh fish and chips. Justin has had some trouble with the chainsaw blade freezing up, and the ice

MADE IN ENG... AN 502/QB
VIEW FROM T...

JUNE 1987

VIEW IN PORCH
AFTER BLIZZARD

14

MADE IN ENGLAND
VIEW FROM THIS SIDE

JULY 1987
REFUELLING IN BLIZZARD

MADE IN ENGLAND
VIEW FROM THIS SIDE

22

JULY 1987

blocks he has just cut refreezing before he has had time to lift them out, but he has come up with some interesting ways of removing them.

24 MAY 1987;
Cornelius van Dorp

It is calm at the moment – clear, starry skies and no wind. The scene is very different from when I last wrote three weeks ago. In between, we had our biggest blizzard yet, lasting 10 days and almost burying our little green box in snow. During the blizzard, the temperature rose drastically and at one stage reached −8°C (18°F). The sea-ice seems to have frozen in for the year in front of the base and with the massive snow drifts everywhere it's a new place: no water, no black volcanic rock, just a vast expanse of white under the moon and stars.

The year is passing rapidly. It's only two-and-a-half weeks to the mid-winter air drop on 10 June. Our field gear is almost prepared, too. The last few days have been spent filling 20 food boxes for sledging trips. The food was carefully worked out in calorific value to last for about 20 person days. We've also got the skidoos up and running and we've been practising along the beach. Only a few more days' work to do on the sledges and we'll be ready to drive over to Scott base to collect our mail and Red Cross parcels.

10 JUNE 1987;
Gudrun Gaudian

We have had a couple of ace blizzards. The first one lasted about 10 days, with gusts of 100 knots – pretty exciting.

When this was over we had literally to dig ourselves out of the base. The front door had completely disappeared behind a wall of drift snow, and we had to use the emergency hatch until Kevin freed the door. Kevin and I filled the snow melter – we had almost run out of water because we hadn't been able to fill it during the blizzard – and it took a day for the water to return to its maximum level.

Then the blizzard started up again. This time the gusts were not so wild, "only" up to 80 knots. Although air temperatures were comparatively high – about −10°C (10°F) – the winds were fierce, and snow was swirling around all over. It took one's breath away.

Cornelius has moved out for a few days as part of his study on the pineal gland.

14 JUNE 1987; Justin Farrely

We certainly seem to be at the centre of a storm: New Zealand and the US have refused to carry our mail. As you can imagine we are very annoyed, not so much at the lack of mail, but at the insinuations that we are not a "fully independent"

expedition, and that the main reason for the drop this week was to provide us with logistical support for the base.

Cornelius arrived back at the base a couple of days ago, having finished his study on himself. He felt exactly the same as I did when I arrived back from my little holiday away: a greater awareness of the comfort and warmth of our little home. A day or so later, he and I went for a walk under the full moon to look at a possible route to take the skidoos to the firm sea-ice at the rear of the base. I hardly recognized the routes I am used to walking, as the wind has scoured the landscape and transformed the scenery.

23 JUNE 1987; Kevin Congalen

Sorry to say that not much of our mail managed to slip through the US Navy sort-out in New Zealand. It appears that "they" don't like some of our photographs of their McMurdo waste dump. It's a shame they don't spend the same amount of time cleaning up their dump as they did going through our mail.

Cornelius has made 75 flags, using yellow, green and red material, which we will use to mark safe routes over glaciers and on the sea-ice. At the moment the bay is open water as the ice was blown out again in the last storm. We won't be able to do a lot of travelling until the bay has frozen back over.

28 JUNE 1987: Cornelius van Dorp

The week started extremely well with our mid-winter dinner on the 21st. You'll not blame us when I say the few days after the dinner were a bit slack. Sleeping and reading were the order of the days – and eating turkey. Incredibly, it took only 36 hours to demolish every last scrap. Then we all slept through another two-day storm. This one gusted to only 60 knots but the sound of the wind shrieking in the rigging still seemed like a good excuse to stay in the bunk and start another book.

Eventually the wind dropped to 10 knots or so at which point everyone emerged from their cabins looking hungry. The ice was blown away yet again, so Gudrun went down to set her fish trap. Kevin almost found himself going for the mid-winter swim he and Justin had so narrowly escaped, as the fish trap started floating out on its own. It was recovered but the following day it had to be chipped out of the ice, so I guess it's back to the tripod and hole in the ice puzzle, when the weather gets better, that is.

Gudrun still seems set on freezing her lungs. This week she "ran the world" for Sport Aid, which involved a 1km (½ml) run up Windvane Hill and back. She set off when the temperature was −27°C (−16°F); she was photographed and videoed from all angles. The BBC World Service was very interested, and spoke to her for quite a while.

The atmosphere is very relaxed and we are all very happy here at the moment.

20 JULY 1987: Justin Farrely

Kevin, Cornelius and I are safely back from our trip, and boy, what a trip. We loaded up a sledge, and hauled it overland to our kick-off point across the ice behind Cape Evans. Using a skidoo, the three of us managed most of the overland route on snow. We had only to unload the sledge once, where we had to lift it across some gravel. Complete with fuel, tents, food, cooking utensils, groundsheets, ropes and emergency equipment, the sledge weighed in the vicinity of 453kg (1,000lb). It is made of wood, with Kevlar shod runners, so that it travels smoothly over the snow.

Once we had deposited all our equipment, including a portable field radio, and our individual packs on the edge of the ice, we headed back for a final night's rest. After a big bacon and egg breakfast the next morning, we hitched up the sledge, and departed on to the ice: the wind was travelling at 5 knots and the temperature was −28°C (−18°F) or so. I drove the sledge, while Kevin towed, and Cornelius rode behind on another skidoo, checking on how we were doing. What an amazing experience: the moon flood-lit the ice and charted our progress; the skidoos thrummed away; and the sledge runners hissed on the ice. It was magical to see the runners conforming to the contours of the ice by flexing over every single mound and hollow.

The ice surface was mostly far from smooth, peppered with small spindrifts of hard snow up to 30cm (1ft) high. We tried to avoid the larger ones to reduce the risk of tipping the sledge. Where the surface was smooth ice or hard snow, the skidoos could manage 25kmph (16mph), but otherwise we were forced to travel at less than walking pace. After leaving the rear of the cape (there was still open water in front of the base), we headed for the Erebus ice tongue.

To cross the tongue we put the second skidoo on a tow rope behind the sledge, while I ran along the side of the sledge, holding on. The tongue has many deep, wide crevasses, and by travelling this way if either of the skidoos or the sledge dropped in one, there would be at least two remaining "anchors" to prevent the loss of equipment. Kevin was invaluable during this part of the operation, recognizing the snow bridges over the crevasses, and guiding us through with no problem at all. Once safely off the tongue, we made a bee-line towards Hut Point Peninsula, and McMurdo base. About an hour later we were greeted by the awesome sight of McMurdo's eerie orange sodium lights, along with the steam issuing from vents in the base complex.

We then travelled on past Scott base, towards our destination: the snow fields near a ski/survival chalet that is cooperatively run by Scott and McMurdo bases. The winds rose to about 30 knots, and that, coupled with a temperature now at −35°C (−31°F) provided an effective (wind chill) temperature of −70°C

(−94°F). When we reached a site that Kevin was happy with, he and I started to dig corner holes straight away for one of our two polar tents.

The next day we were treated to a magnificent sight: six men from Scott base walked past in special harnesses, towing a sledge. They were pleased to see us, and we chatted to them as they hauled the sledge to the top of the hill behind the chalet. It was good to meet some other people, the first in a very long time it seemed. (It was, in fact, five months.) Later on, we met some of the American contingent and had some equally warm and friendly chats with them. It was really nice to know that the individuals at these bases are interested and concerned, not only in our welfare, but in the welfare of the continent as a whole.

On our return journey, we took only the skidoos, leaving our sledge, a lot of equipment and our tents ready for our next journey that way. It took us only one-and-a-half hours to get home compared to four hours to get there.

1 AUGUST 1987: Kevin Congalen

This week started off with a celebration for Gudrun's birthday. So we had a relaxing day on Sunday. The rest of the week our little green base was turned into a hive of activity. We started off by building ourselves a small hut out of plywood, 2m² (7ft²), complete with a small round window, cooking bench, vents, and bunks for two. The hut is mounted on skids so that it can be towed by our skidoos. It will be used for shelter by Gudrun when she is on the sea-ice, carrying out her fish studies. We have called it Little Greenpeace 3.

8 AUGUST 1987: Cornelius van Dorp

The last 10 days have seen an amazing change in the amount of light we receive. Ten days ago we started getting some rich golds and reds in the northern sky around midday. After three months of darkness these colours came with an almost emotional force. Now, only a week later, we have six hours of twilight every day.

17 AUGUST 1987;
Justin Farrely

The biggest excitement was Kevin and Gudrun departing on a five-day trip, with our two skidoos and a whole lot of other equipment. They will travel past Scott base to our original campsite to check that our tents, equipment and sledge are safe. With the amount of daylight that there is now, it is practical to travel during the "light" and camp during the "dark". Although they expect to be away for five days, it could easily turn into 20 if the weather packs in.

It is snowing outside the base at the moment; the snow could be used as baby powder it is so fine. Visibility is about 10m (35ft). I shall not be wandering outside, because, silly as this may sound, it is quite possible to get lost travelling even short distances in these conditions. You walk out, and the next thing you know the whole world is white. Often, the only way of telling that you've arrived at a new location is when your nose gets flattened when you smack into an object. Worse, of course, is not walking into something when you were expecting to.

22 AUGUST 1987;
Cornelius van Dorp

This should have been the week of the first sunrise. (In 1911, when Scott's party was here, the sun rose on 21 August.) This year, the day started out clear with a pale blue sky; Mount Erebus loomed over the cliffs and a pink plume of smoke drifted from the crater. But then, later in the day, in came the cloud, especially thick over the Barne Glacier to the north. So we have still to experience the thrill of that first sight of the sun.

31 AUGUST 1987;
Gudrun Gaudian

Kevin and I finally began our epic trip. The weather cleared and, although we hadn't slept for 24 hours, we packed the skidoos and dressed in layers of warm, protective clothing. Ahhh, what an ace feeling it was, to drive across the sea-ice towards the red-tinged mountains of Victoria Land. It was −32°C (−26°F) with little wind and a clear sky. After about an hour, dodging snowdrifts and bumps of ice, passing pieces of rubbish from the Mactown garbage dump that had blown out, we rounded Hut Point Peninsula and saw the lights of McMurdo base.

I had to stop my skidoo, "Ishtar", and stare. So many lights. Finally, we reached the tent by the ski chalet. It was dark, the stars were out, and the aurora australis looked impressive. We had just cleared the inside of the tent of snow, when we heard a vehicle approaching the chalet. It contained the first new faces I'd seen for months and I couldn't stop staring.

The weather turned bad and we spent several days in the tent, only going out to check the sky. At last the weather cleared sufficiently for us to take both skidoos to Scott base post office. While I posted the letters, we chatted to some of the Scott base personnel. I was showing off, proclaiming that "Ishtar" started so easily because of "the woman's touch". I wish I'd kept my mouth shut.

I tried to get "Ishtar" from neutral into first gear, a difficult exercise with thick gloves on, and she suddenly took off at 60kmph (37mph), full throttle, up the hill. I just managed to dive off before she buried herself beneath Scott base kitchen, stuck under one of the cross-beams of the construction. I saw our travelling programme for the summer dwindle to nothing. The lads from Scott base came out and, after having taken the obligatory photos, gave us a hand heaving "Ishtar" back into the open.

We wanted to check "Ishtar" out thoroughly, so we decided to drive home the next day, weather permitting. Just as we passed the pressure ridges near Scott base, Kevin's skidoo turned over.

He did not seem too bad, although his shoulder was hurting. We turned his skidoo upright again and tried to get it started. We failed. We secured the sledge and drove back to camp on trustworthy, battered "Ishtar". There we examined Kevin's shoulder. He was now in a lot of pain and even a strong painkiller didn't make much impact.

Finally the weather cleared again. Kevin's arm had improved a lot – it was resting in a sling now – and I looked a little livelier, so we decided to make for home. We drove back into the sunset. It was the first time we had seen the sun for months. What an exhilarating moment. We picked up Kevin's skidoo and, with a lot of tricks and cunning, got it started. Kevin could drive OK, and he pulled the sledge. We arrived late dusk.

19 SEPTEMBER 1987;
Cornelius van Dorp

The warm, windy weather continues, and in front of the base there is open water as far as the eye can see. A few days ago the temperature rose to −9°C (16°F), while the wind gusted to 80 knots. Today, however, we had a rare still day at −14°C (7°F). The rest of the 11 Scott base winterers visited. They were all very impressed and took heaps of photos of World Park base and the open water.

The sun rose fully on 25 August but we felt the real effects on 24 August, when it produced a brilliant golden glow over the Barne Glacier.

Soon after the invalids came limping home on their battered skidoos, Justin and I set out on a mail run to Scott base and Mactown. We too were beset with foul weather, spending several days comfortably cocooned in our down bags, while outside the tent the blizzard shrieked and the snow swirled.

Eventually the storm subsided and we set out for home, slipping and sliding over the snow hummocks, skidoo humming and runners hissing, our rainbow sledge loaded down with mail.

We've already had some Weddell seals and one column of emperors marching towards Hut Point. All the signs are that summer is almost upon us in the amazing white south.

The first rays of the spring sun

INTERVIEWS

Key figures in the Greenpeace Antarctic campaign, Kelly Rigg, Roger Wilson and Jim Barnes summarize recent developments in the region and the way Greenpeace has influenced – and continues to influence – them.

KELLY RIGG

Based in Washington DC, Kelly Rigg is the current coordinator for the Greenpeace Antarctic campaign. She has worked on Greenpeace campaigns to prevent offshore oil drilling in sensitive marine ecosystems, and has monitored the minerals meetings and the UN debates on Antarctica.

66 People often ask me why Greenpeace has invested so much effort and so many resources to protect Antarctica. After all, they say, Antarctica is remote, relatively unspoiled, and there are many other issues confronting us – the toxic waste crisis and the threat of nuclear war.

The first answer is that Greenpeace works on many fronts, and in fact the problems of toxic waste disposal and nuclear disarmament are among our highest priorities. The second is that Antarctica is under siege.

Up until the last few years, the principles of scientific cooperation were held paramount. Antarctica is the one place in the world where all nations have

agreed that nuclear weapons, for testing or deployment, do not belong. In the 30 years since the Antarctic Treaty was negotiated, however, the worldwide plunder of natural resources has reached crisis proportions. A battle over indiscriminate development versus sustainable use and conservation is now being fought around the world.

Antarctica could become the focus of energy conflicts similar to those seen today in the Persian Gulf

Environmental activists are campaigning to prevent the opening up of pristine wilderness areas to oil and gas drilling, mining, logging and damming, because we realize that there are no such things as purely "localized" effects.

The current crisis in the Persian Gulf is a case in point. It is almost unbelievable that so many nations have directly involved themselves in a conflict between two neighbouring countries simply to ensure a "stable" energy supply. What is stable about that? Surely a more stable supply of energy would include safe, renewable alternatives which can be produced at local levels and the supply of which does not depend on military protection. Now, increasing international interest in Antarctica is often described in terms of the potential future supply of "strategic" minerals. It is not difficult to imagine the consequences of such an attitude.

Developing Antarctica's mineral resources would therefore imperil the foundations of peace in the region by introducing resource competition where cooperation and the sharing of data once prevailed. There can be no assurances that energy conflicts would not be exported to Antarctica, particularly given the likelihood that Antarctic petroleum or minerals will not be developed until

supplies from relatively less expensive sources elsewhere begin to dwindle.

If we can't save Antarctica, what hope do we have of saving the rest of the planet?

If we can't leave this place as it is – when there is no immediate need for its hidden mineral wealth, when there is sufficient time to find renewable alternatives, and when there is at least a degree of international cooperation in Antarctic science – how can we possibly hope to reverse the destructive tide in other areas of the world?

The public is beginning to realize the significance of Antarctica. In the last two years more than a million people have signed the Greenpeace Antarctic Declaration, a document that lays out in Treaty style the principles that would govern a World Park. The Fourth World Wilderness Congress, which met in September 1987, recommended a moratorium on minerals activities at least until the year 2000.

Interest in Antarctica continues at the United Nations. The nations that have initiated the Antarctic debates have questioned the right of the Treaty Parties to make deals over mineral rights.

The response of the Antarctic Treaty Consultative Parties is a defensive one

Rather than listening to the concerns of the dissatisfied non-Treaty states, and responding appropriately, the Treaty states portray the debate as one in which outsiders are attempting to destroy the Antarctic Treaty itself. I don't believe this was or is the intent of the majority of concerned states. In my view, it is the

Consultative Parties who threaten to destroy the Treaty.

The strength of the Antarctic Treaty was the agreement of participating nations to limit and structure their own behaviour in Antarctica. It prohibits activities that might endanger peace and scientific pursuit in the region. The Treaty states agree that Antarctica must not become the scene or object of international discord, and that the use of Antarctica should ensure international harmony in furtherance of the principles of the UN.

The Treaty states should acknowledge that Antarctica **has** become the object of discord, because of their own attempts to take possession of Antarctic resources.

This is not to say that Greenpeace's views are entirely in accord with those of Malaysia and the other debate participants. Developing countries have suggested that Antarctic minerals ought to be considered the Common Heritage of Mankind, which means that if minerals are discovered and extracted there must be an equitable sharing of benefits. As Greenpeace is opposed to **any** Antarctic mineral exploitation, we have not aligned ourselves with this principle.

Greenpeace supplies information about Antarctica to all UN delegations

We do, however, support some role for the UN in Antarctic affairs, in the form of a special office or committee that could give voice to the concerns of non-Treaty states. We hope that along with an increased information flow will come a better understanding of the importance of environmental protection. This is the role we've attempted to play at the UN, by distributing briefing documents to all delegations during the year and immediately prior to each debate.

Much of the ATCPs would like to deny that international and public pressure has resulted in significant changes in the way the Treaty System operates, there are undeniable signs, both positive and negative, that the message is getting through. First, because of UN pressure, the Treaty states have accelerated the pace of the minerals negotiations and hardened their resolve to complete the Convention as soon as possible.

Changes are taking place in the Treaty system, and meetings are more open

UN pressure has also caused the ATCPs to open up the Treaty significantly. The fact that they admitted six new Consultative Parties between 1983 and 1987, when only two had been admitted between the Treaty's inception in 1959 and 1983, is indicative of that. It is also true that the standards for achieving Consultative status have been relaxed recently.

Another result of the UN debates is that the Treaty powers now invite the Non-Consultative Parties to observe Treaty meetings, including the highly secretive minerals negotiations. In addition, in 1987, for the first time the International Union for the Conservation of Nature, SCAR and the World Meteorological Organization were admitted to certain discussions during the Consultative Meeting.

By gaining greater access to information, the Non-Consultative Parties are able to contribute more to Treaty decision-making. However, this has been achieved on the understanding that at the UN debates the Non-Consultative Parties will vote (or "not participate") as a block with the Consultative Parties.

Public opinion has had an impact on the Antarctic Treaty System and on Antarctic policies

Pressure from the public in support of the World Park option in general, and of more specific environmental protection measures, has led to slower, but nonetheless identifiable changes. The 1987 Consultative Meeting saw more topics of environmental importance on the agenda than ever before. These topics included waste disposal, the need for environmental impact assessment, new kinds of protected areas, and safety measures for scientific drilling.

In the United States, Greenpeace and other Non-Governmental Organizations (NGOs) are taking a close look at the environmental practices of the US Antarctic Program. Waste disposal is a particular focus. The US programme is now being reviewed, and clean-up efforts may begin this season.

Unfortunately, however, at all meetings of the Antarctic Treaty System the pace of decision-making has been glacial at best. For example, the problem of waste disposal was brought up at the 1985 Consultative Meeting and NGOs were disappointed by the fact that rather than take action, the ATCPs referred the issue to SCAR for further study. The SCAR study was due out in June 1987, in time for consideration at the October 1987 Consultative Meeting. Unfortunately, SCAR was unable to produce the report, due to the lack of information provided by member states, and the issue was once again deferred.

There is a movement towards a World Park among some of the Treaty states

In the minerals negotiations it has been difficult to achieve an agreement that accommodates the divergent interests of the various factions – claimants, non-claimants and developing countries. In this case as well, the environment has suffered as a result. Many important environmental issues are being swept under the rug in an effort to achieve political compromises.

However, at the last minerals meeting in May 1987, there was at least a glimmer of hope. Ambassador Zegers of Chile stood up during one of the sessions and said that given the way things were going, there would be little left in the Convention to accommodate the interests of Chile and other claimant states. He suggested that the ATCPs might as well negotiate a World Park instead. In later discussions with NGOs, he revised his statement to read "Antarctic Treaty Park" and agreed to his statement being quoted in the Greenpeace newsletter.

Many delegated to Antarctic Treaty Meetings privately express their discomfort with the exploitative direction the Antarctic Treaty System has taken. Some provide us with secret documents in the hope of increasing public attention and pressure. Others simply provide valuable advice. Other delegates seem intent on waging a disinformation campaign about our motives and activities. It is my fervent hope that those of the first group will eventually succeed in setting national policies. 99

ROGER WILSON

Roger Wilson has been the prime mover of Greenpeace's Antarctic campaign. He is now concentrating his efforts on the political aspects of the campaign.

In June 1982 I was living in Wellington, New Zealand, when the first of a series of negotiating meetings on Antarctic minerals took place there. Along with a number of other people who were interested, not only in Antarctica, but in conservation politics generally, we began a lobbying operation at that meeting.

After the meeting it was clear that the issue was not going to go away. Somehow we had to ensure that there was a continued environmental influence at these negotiations, and that it was organized internationally.

I'd worked with Greenpeace in New Zealand previously, although I wasn't actually employed by them at the time. I concluded that Greenpeace was the vehicle through which we could most effectively monitor and influence the negotiating process. Greenpeace had the network that was needed in order to operate internationally, it had a high media profile; and it was respected around the world.

I didn't know it at the time but Jim Barnes in Washington was also thinking along the same lines and was also putting suggestions to Greenpeace to the effect that Greenpeace should adopt Antarctica as a major campaign issue.

Jim was to visit Australia in October 1982 and I went across to meet him. Together, we wrote a proposal for Greenpeace International. Greenpeace was interested but not willing to make an immediate decision on a major new campaign area.

After the second negotiating session in January 1983, which was also held in Wellington, the situation began to appear much more urgent. I travelled to England in May 1983 to try to persuade the Greenpeace Council of the need to become involved as quickly as possible. Jim and I put together a new proposal for that meeting in June 1983. We were successful and Greenpeace agreed to adopt Antarctica as a campaign issue.

Even at that stage, the aim was primarily to keep track of what was happening on the minerals negotiations. We were looking at the issue from a political viewpoint, rather than any other perspective. That was, after all, where most immediate threats to the future of the continent lay.

Greenpeace adopted and developed the concept of a "World Park"

Obviously, very quickly after that, we had to develop our "World Park" strategy. The term "World Park" was invented by the New Zealand Government in the 1970s. The implication was that mineral activity in Antarctica would not be permitted, and that the continent would continue to be the preserve of scientists undertaking non-damaging research.

Because the concept had once been promoted by a government, it had gained a certain cachet. The catch-phrase "World Park" was too good to lose. So we used it in the early years of the campaign, but without having a real sense of what we intended by it other than that, in general terms, we should conserve Antarctica rather than exploit it.

Later, in 1985, we realized that we had to be more specific about what the "World Park" phrase actually meant. We derived a set of four principles which we enshrined in our policy. We deliberately made no attempt to say *exactly* how we would implement the World Park concept. If we could make those principles the basis of all Antarctic decision-making, we would

consider that we had achieved the World Park concept. We didn't mind by what political mechanisms it was done. It could be within the Antarctic Treaty System or it could be outside it.

It was important to become more familiar with the nature of the meetings concerning Antarctica

Immediately after the Greenpeace decision to take on the campaign in July 1983, there was another in the series of minerals negotiations in Bonn. Jim and I went across to Germany, and our German office supplied a team of volunteers to help at that meeting. There were also representatives of other NGOs present, from as far away as Australia. But in terms of the amount of research needed to run a full-scale campaign, we were still flying by the seats of our pants.

In August of that year I went to a CCAMLR meeting in Hobart, Tasmania, and from there to the Antarctic Treaty Consultative Meeting in Canberra to try to get a feel for the two other political forums within the Treaty system. Both of them were important, but in neither case did we know very much about them in practice.

We turned our attention to raising public awareness primarily through direct action

I'm very much aware that successful environmental campaigns have always had three components: the political component, the public awareness-raising component and the scientific component. I've never seen a really successful campaign that hasn't combined all three. So when we actually set up the campaign properly, it was important to get working in each of those three areas.

Direct actions are usually oriented towards raising public awareness. We had taken direct action in December 1982, even before the international campaign started, against a Japanese seismic survey vessel which berthed at Lyttelton, New Zealand. It was a sort of dummy-run, though it was very amateurish.

People attached themselves to the mooring lines of the boat and tried to delay its departure. Naturally, it didn't delay them substantially but it made front page news in newspapers throughout New Zealand and was quite a boost to people's ideas about having an Antarctic campaign – in New Zealand, at least. We mounted another direct action later against the same vessel.

In New Zealand, of course, Antarctic issues are very much more widely explored in the media than they are in almost any other country. Nobody had ever taken any direct action against Antarctic-bound boats or anything like that previously. It had always been a purely scientific issue and nobody had thought about the political implications of the research that was going on there. So, to an extent, we were breaking new ground.

Greenpeace alerted the Antarctic Treaty members to the building of the French airstrip

In between the two minerals meetings in Wellington – around October 1982 – I was passed some information by a very senior civil servant in New Zealand who had had private correspondence with a counterpart in the French civil service. This counterpart was personally very concerned about the proposed French airstrip development at Dumont d'Urville and predicted that it would cause severe damage to the local fauna and flora. My civil servant obviously didn't want to be associated with this but thought that somebody should do something about it.

I'm really indebted to this person for getting us started on this issue. Considering his position now, it was quite remarkable. The same thing probably wouldn't happen today.

We investigated and found that the basis of the information was correct, so we put the information in front of the delegates at the second minerals meeting in January 1983. The delegates at the meeting didn't actually believe us. They assured us that the French would have told them if they were undertaking such an activity on that scale. It just wasn't possible, they said.

It was quite a coup for us in that we had alerted the Antarctic Treaty System to something that was going on that they should have known about. That was the start of it. We continued to bring the issue

PENGUIN PROTESTERS *Greenpeace draws attention to the French airstrip.*

up at every subsequent meeting of the Antarctic Treaty, whether it dealt with minerals or whether it was a meeting of the Consultative Parties.

The airstrip was a significant issue for a number of reasons. The first and most important reason was that it represented very serious allegations of breaches of the Agreed Measures by one of the member nations of the Antarctic Treaty. Second, it was an eye-opener in that the rest of the Treaty states were patently unwilling to address it. They found it embarrassing even to think that one of their number was not abiding by the regulations mutually agreed. The Antarctic Treaty member states pride themselves on their unity.

Photographs of the destruction backed up our claims

Though ours was a serious allegation, they still wouldn't have investigated it but for the fact that we were able to find a French scientist who was prepared to give us film footage and still photographs of the construction programme, showing penguins lying injured and bleeding, which we put in front of delegates at the fourth minerals meeting in Washington in January 1984.

The mineral negotiations played a large part in the early stages of our campaign

because they were the only meetings where we had access to the decision makers when they were all gathered together in one place.

The response was mixed. Some people were concerned; others went so far as to say that those photos "could have been taken anywhere", implying that we may have been using the photos dishonestly to embarrass the Treaty system. At that stage also, the film footage was shown on television in Britain, France, the United States, New Zealand and Australia.

The French airstrip issue has been one of the most important we've dealt with. While, in practical terms, the number of birds killed or injured was very small, the scale of the project was vast. Further, they'd done no Environmental Impact Assessment until after the project had actually started.

We turned our attention to organizing an expedition to Antarctica to set up our own base

It had always been at the back of our minds that once the campaign was established we would need to go to Antarctica ourselves. This idea developed during 1984, and it was at the Greenpeace Council meeting that year that the budget was approved for the expedition.

173

The first thing to do was hire people to plan the logistical side, and we began looking for such people at the end of 1984. Our first employee on the expedition started in early 1985, and during that year we made all the basic preparations. These involved obtaining a boat, gathering all the materials and equipment together, and getting a base ready for our departure from Europe, which turned out to be early August 1985. The boat left early as a result of the sinking of the *Rainbow Warrior*, so that it could go to Moruroa *en route*.

We found a basic design for the Antarctic base, and adapted it to our own needs. The company concerned designed it and pre-fabricated it and it was all packed away and loaded on to our boat in Hamburg so that we could re-erect it on arrival in Antarctica.

Pressure of time was the main problem. There were just so many things to attend to: getting the base together, buying a helicopter, setting up satellite communications, finding the personnel. Finding the right personnel was probably the hardest. The sort of personnel we needed did not grow on trees. Greenpeace's budget was absolutely minuscule compared to a nation's budget for establishing an Antarctic base. We worked on an absolute shoestring.

On our first expedition, weather conditions were against us

The first year we went down there – the 1985/86 southern summer – we faced the worst sea-ice for 30 years. It was not possible for us to get within range of our objective: Cape Evans on Ross Island. We were able, however, to make a landfall at the Bay of Whales on the Ross Ice Shelf – an "icefall", I'd guess you'd call it. This was the place from which Amundsen set off when he headed for the Pole. It was the only landfall that we made.

When it became quite clear that we weren't going to get sufficiently close to Ross Island to undertake the erection of the base, we had to make some very hard decisions. I don't like to use the word defeat but we had to accept that if we weren't going to be able to establish the base that time, we would have to return to New Zealand. So we did.

When I say "we", it's hard for me to imagine that I wasn't personally aboard the ship, I felt so close to them. However, I was back in England, speaking on the phone every day to them, trying to find ways in which the situation could be salvaged. The ultimate responsibilities were mine, and it was one of the most stressful times of my life.

Some people doubted that we would ever reach Ross Island

As a result of our turning back, there were people who questioned whether we could ever guarantee to get in there. There were also the doubters, who had heard stories of poor morale on the boat. Morale, as a result of sitting down there for weeks and weeks doing nothing, was very low. There were all sorts of tensions on board which inevitably were blown up by the isolation. These sorts of things did influence a lot of people and there were moves to cancel the expedition part of the campaign or to modify it drastically.

However, the Greenpeace Council had already budgeted money for a resupply of the base prior to the first expedition leaving. They set the budget for 1986 during 1985, so we had the money and we had a mandate to get on and do the job – so we went ahead, and we did it in the 1986/87 southern summer.

This time, we used a larger helicopter, which meant that we could, if necessary, move the entire cargo from 30km (19mls) out. As it turned out, we got to within 200m (220yd), so it was used primarily for very short hops. But it was really invaluable and if you had to find a single factor that made the difference between this year and the last, the helicopter was it.

The establishment of a base meets the three criteria of our campaign

We have now established the first permanent, non-government base in Antarctica. It is not certain that we will occupy it ourselves indefinitely. If we found some other organization that was clearly going to follow an environmental mandate then we might at some stage consider sharing the use of it, or even letting them use it for a while. We don't have any firm plans other than that it should be used for environmental purposes and, because it's such a large investment, it has to stay there for some length of time to earn its keep.

Politically, by establishing a base we have fulfilled the same kinds of criteria that are applied to nations that seek membership of the Antarctic Treaty as Consultative Parties. It's our contention, having established a base, staffing it through a winter and conducting a scientific programme, that we, at the very least, deserve observer status in the various institutions of the Antarctic Treaty System. We have demonstrated an interest towards and concern for the Antarctic that is the equal of some of the smaller countries involved in the Antarctic Treaty System. Some have done very little to justify their consultative status.

From a public awareness point of view, the base is a platform from which to demonstrate to the world that there is a problem in Antarctica that needs to be addressed. It's a platform from which we hope we'll be able to make people more aware of the issues and the hardships that are involved in being there.

Scientifically, we have a commitment to maintaining Antarctica as a zone of peace for the use of scientists. We want to demonstrate some kind of solidarity with our scientific colleagues, many of whom clearly recognize the problems that are emerging in Antarctica, and have been extremely helpful. So we're undertaking a small scientific programme.

Greenpeace has become a force to be reckoned with in Antarctica

All in all, I believe we have created a good, well balanced campaign that has the potential to make a significant impact on Antarctic developments. It's a long, slow process, however – the decision-making process is glacial. But we're in there now, and the decision makers recognize us as a force to be reckoned with. Even some of our critics give us implicit encouragement in our work. I think they feel in some way secure in that they know we won't let them get away with anything outrageous, and I think they're actually scared of us. As far as I'm concerned, **"** that's just fine.

JIM BARNES

Jim Barnes was operating as a public interest lawyer at the Center for Law and Social Policy in Washington DC when he formed the Antarctic and Southern Oceans Coalition in 1977 to try to raise international awareness on the issue. He is currently Director of The Antarctica Project, and counsel to ASOC.

"In 1977, several people in the United States, with many contacts overseas, wrote a circular letter to about 25 organizations around the world, including major groups in England, Australia, New Zealand, Canada and the US. It was proposed that a citizens' coalition be set up to monitor what the governments were doing, to report to each other what they found out, and to try to work together to influence the positions of our individual governments. We wanted to see if there was enough interest among citizens to have a campaign on this whole question of protecting Antarctica.

It was agreed that a World Park petition would be circulated among the public. We drew the petition up jointly, and got several hundred thousand signatures on it between 1977 and 1982. They were officially transmitted to the Secretary General of the UN in 1983 in connection with the first debate at the General Assembly on the issue of Antarctica.

Non-government environmentalists saw the UN as a good forum in which to raise this issue, to let a lot more nations know what was going on. Because of the debate during the last few years at the UN, every nation in the world now knows quite a lot about Antarctica.

Greenpeace has an international structure and could mount a worldwide campaign

I'm a global environmentalist and I support all environmental groups. They may have different approaches and different constituencies but if you put them together, then the environmental community has the most effective possibility for accomplishing its objectives.

So, starting in 1981/82, consistent with the position of Greenpeace as a protection and action-oriented organization, it was proposed that there be a campaign to take the World Park concept forward in a political way, using the media and direct actions to dramatize what was going on.

We wanted to work on the broadest possible front

We took a very firm protection position, and it was agreed in the initial stages that advantage would be taken of any potential forum. If that included the United Nations or the Non-Aligned Movement, fine. If it included any part of the Antarctic Treaty System, fine. We would take all routes to inform the public and lobby decision makers on the World Park concept.

Then we came up with the idea of writing a new document – the Antarctic Declaration – which is the same general idea as the World Park petition but it puts it into the Treaty context. It takes the language of the Antarctic Treaty itself and proposes certain modifications that would carry forward the World Park idea.

The participation of Greenpeace in this UN debate has been absolutely essential to any success achieved there. The annual briefing documents, which Greenpeace has published for the last four years, have been very important and widely used by government representatives at the UN.

Progressively Greenpeace laid out many interesting ideas and critiques, ranging from how the fisheries ought to be operated, to the need for an Antarctic Environmental Protection Agency; from critiques of the minerals regime as it was evolving, to critiques of the airstrip that the French were building.

The debate at the UN will continue to be important

Unfortunately, in 1985 the Antarctic Treaty Parties decided that they would stop participating in the debate. So now there's a dichotomy at the United Nations between those nations that are in the Treaty and those that are not.

We would like to see a consensus restored at the UN because it's not a very productive situation. On the other hand, it certainly has been very interesting to see what happens as a result of the pressure on the Antarctic Treaty Parties created by the debate at the UN.

Greenpeace and ASOC would like to see the UN set up a professional office on Antarctica, a non-political committee which would prepare reports and studies and be a forum for the exchange of views and information about the Antarctic region, so that all countries, whether or not they are members of the Treaty system, would have some place to express their views and obtain information.

The minerals prospecting moratorium is being only loosely respected

The minerals convention negotiations have been going on formally since 1982 and informally since 1980. It's a long-term effort. That's perfectly fine with us because environmentalists can gather their strength only slowly, especially as convention documents are secret. As the worldwide campaign grows, NGOs have more of a chance to influence the positions of individual governments.

There's a very strong internal willpower, you might call it, within the Antarctic Treaty group to complete the minerals negotiation. They know that if they don't, it will be seen as a failure and may invite greater attention by the UN.

If a minerals regime is brought into force, prospecting activities – mostly geophysical seismic work – would probably proceed without many controls.

The present moratorium is not binding and therefore governments wink at it. I'm pretty certain that countries like Japan, Germany, the Soviet Union and maybe France are actually carrying out what could be called prospecting activities right now. With a convention in force, any country would have a right to carry out these activities.

We have to block exploration for minerals from the beginning

If negotiations were concluded, a meeting of all members of the minerals convention would have to agree *by consensus* to open an area to potential minerals activity. That would be a good opportunity for Greenpeace and other organizations around the world to lobby to block development. If we can't convince one country to block minerals development then we're not very good lobbyists.

If an area were opened, could an operator or a state party get a licence to explore? There would be some pretty significant hurdles. An operator would have to bring forward adequate information on which to base a rational decision about whether to allow exploration. The scientific advisory committee would then decide whether the information brought forward by a potential operator were sufficient. This would be another opportunity to try to block any further activity.

Activity would be best blocked at the beginning because the decision to open an area would require a consensus; thereafter it would probably require only a high majority to approve certain actions like giving out a licence.

The implementation of CCAMLR has proved difficult

In the early days of ASOC, one of the first things that we ever did as a group was to convene an international panel of scientists, starting in 1979, to give advice to the governments about what should be in the fishing agreement.

ASOC lobbied a number of key countries to insist on an "ecosystem as a whole" approach in the draft and the final conventions.

NGOs did a pretty good job in the entire convention but it proved impossible to convince the governments to adopt a decision-making approach that required less than consensus.

The main reason for this failure was that the claimant states, of course, wanted to make certain that they could block anything they didn't like, while the fishing states, primarily the Soviet Union and Japan, were very interested in blocking conservation measures. So there was an alliance of interest between the sovereign claims countries, including the UK, Australia and New Zealand, and the fishing states.

In practical terms, fishery is only as regulated as the governments will agree to make it and so the fishing is basically allowed until it's curtailed. That's why there has been a lot of trouble with the implementation of this convention.

In spite of that, I do think the advisory scientific committee to CCAMLR is pretty good. It has some very good scientists on it. They put out excellent reports, which are increasingly difficult for governments to ignore.

The Antarctic ecosystem may become the first to be truly managed

Environmentalists are also trying to work with the ecosystem modelling group under CCAMLR. This is a very innovative idea that only exists in the Antarctic. The group comprises scientists from a number of disciplines. They are trying to develop models for the ecosystem to tell what indicator species ought to be monitored in order to reveal what's happening to the ecosystem as a whole.

To give you an example: the back teeth – the molars – of certain seals are like trees – they have rings. Analysed over a period of years, they let you know what the growth rate of the seals has been in response to food. If food is in short supply in a particular area, maybe because of over-fishing or environmental perturbation, then those teeth will have narrower rings. You can also monitor things like age at first pregnancy of certain species and the weight of the litter.

They have a pooled data system and, if NGOs can continue to strengthen the

willpower of key governments and persuade them to put up more money for this research and data analysis, their work will lead to true ecosystem management in the Antarctic.

Only modest changes are likely to be made to the Treaty system when it is reviewed

In 1991, for the first time since its inception, an individual country or group of countries can propose that changes in the structure or functioning of the Treaty system be made.

Now obviously, in a system that has as many different interests to satisfy as the Antarctic system does, it will be fairly difficult to convince very many countries to go along with any major changes. That suggests that a review conference will propose that only modest, evolutionary changes be implemented.

It might recommend: that a secretariat be set up for the Treaty system; that some form of infractions committee be set up; and perhaps that an Environmental Protection Agency for the region be considered, although this is unlikely.

I think there will be more and more protected areas established in the Antarctic, and of a new sort. There'll be large areas for protection of habitat and wilderness, analogous to conventional national parks and protected areas, as opposed to the very small-scale protected areas that have been agreed on so far.

So, in short, I don't look for any dramatic departures in 1991. This is an evolving system. It's got a long way to go but it has interesting potential and Greenpeace, along with a number of other organizations, is playing a major role in it.

As the Greenpeace campaign evolves, there will be increasing ways for Greenpeace to lobby the governments, both at a national level and as a group at the international meetings.

You have to look at these things over the long term. The campaign that has been put together over the last few years has been really successful and it's had a major impact on what the governments are doing. It will continue to have that kind of impact in the future. **99**

APPENDICES

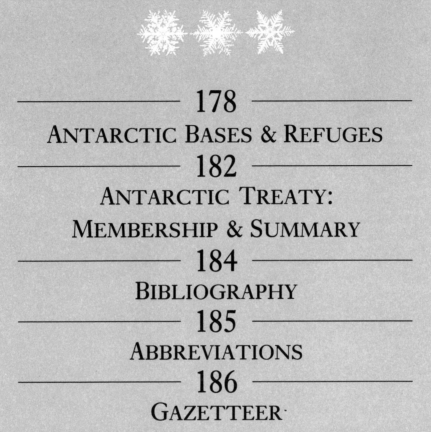

ANTARCTIC BASES & REFUGES

The following tables and map identify the status, location and scientific programme of every base in Antarctica.

ARGENTINA

BASE NAME	LOCATION	DATE OF INSTAL.	POP. sum.	win.	SCIENCE
ALL YEAR ROUND					
Belgrano II* (Army)	77°52'S 34°34'W	5.2.79	19	18	meteorology, atmospheric physics, geomagnetism, ionospherics, physiology
Orcadas (Navy)	60°45'S 44°43'W	22.2.04	49+	14	meteorology, atmospheric physics, geomagnetism, glaciology, biology
Esperanza (Army)	63°24'S 56°59'W	31.3.52	50+	29**	meteorology, atmospheric physics, glaciology, geology, geophysics
Marambio (Air Force)	64°13'S 56°38'W	29.10.69	100+	42	meteorology, atmospheric physics, glaciology, geology, geophysics
San Martin (Army)	68°07'S 67°07'W	29.3.51	28	15	meteorology, atmospheric physics, ionospherics, geomagnetism, physiology
Jubany (I.A.A.)**	62°14'S 58°38'W	12.11.53	17	11	meteorology, atmospheric physics, biology
SUMMER ONLY					
Primavera (Army)	64°09'S 60°57'W	8.3.77	8		meteorology, biology, geology (occasional)
Teniente Matienzo (Air Force)	64°58'S 60°64'W	15.3.61	7		meteorology, solar physics
OCCASIONAL*					
Melchior (Navy)	64°20'S 62°59'W	31.3.47			
Almirante Brown (Army)	64°53'S 62°53'W	6.4.51 (partially destroyed by fire, 1984)			
Petrel (Navy)	63°28'S 56°17'W	22.2.67			
Decepcion (Navy)	62°55'S 60°46'W	25.1.48			
Camara (Navy)	62°36'S 59°54'W	1.1.53			

*Belgrano II, near the Filchner Ice Shelf, replaced Belgrano I, built in 1955 and abandoned in Jan 1980 after succumbing to the pressures of ice.
** Instituto Antartico Argentino – a civilian agency.
*** Closed for several seasons; may be reopened or occasionally used in the future.

AUSTRALIA

BASE NAME	LOCATION	DATE OF INSTAL.	POP. sum.	win.	SCIENCE
ALL YEAR ROUND					
Mawson	67°36'S 62°52'W	13.2.54	36	28	meteorology, upper atmospheric physics, geomagnetism, geology, glaciology, marine biology, cosmic ray physics, pollution studies, human physiology
Davis	68°35'S 77°58'W	13.1.57*	31	24	meteorology, upper atmospheric physics, geomagnetism, geology, glaciology, terrestrial and marine biology, human physiology
Casey	66°17'S 110°32'E	Feb 69	38	32	ionospherics, meteorology, geomagnetism, geology, glaciology, marine biology, human physiology, environmental studies
SUMMER ONLY					
Edgeworth David (Bunger Hills)**	65°51'S 100°30'E	14.1.86	22		geology, geomorphology, terrestrial biology
Commonwealth Bay	67°00'S 142°43'E	Dec 77	7		supports restoration of Mawson's hut; party contains architect, surveyors and archaeologist – also geology, biology and glaciology
SUB-ANTARCTIC – ALL YEAR ROUND					
Macquarie Island	54°30'S 158°56'E		26	20	meteorology, upper atmospheric studies, geomagnetism, ozone, solar radiation, seismology, terrestrial and marine biology

AUSTRALIA cont.

* Closed from January 1965 to February 1969.
** Heavy pack-ice off the Shackleton Ice Shelf prevented Nella Dan from establishing Edgeworth David in 1985/86, causing a postponement of the Bunger Hills programmes.
Note: In 1986/87, summer programmes were conducted on the Law Dome (67°S 112°E, geology, terrestrial biology), the Larsemann Hills (located at 69°24'S 76°23'E (1986/87 season) with 22 geologists and biologists), the Scullin Monolith (67°47'S 66°54'E, 150 kms east of Mawson, ornithology and geology), and the sub-Antarctic Heard Island (geology and biology).

BRAZIL

BASE NAME	LOCATION	DATE OF INSTAL.	POP. sum.	win.	SCIENCE
ALL YEAR ROUND					
Commandante Ferraz	62°05'S 58°20'W	6.2.84	40	12	meteorology, upper atmosphere physics, geomagnetics, glaciology, biology, human physiology
SUMMER ONLY					
Astronomo Cruls (Nelson Island)	62°14'S 59°00'W	Jan 85	6		geology, mapping
Engenheiro Wiltgen (Elephant Island)	61°13'S 55°21'W	Jan 85	6		geology, sedimentology, avian biology
REFUGES					
Padre Balduino Rambo	62°11'S 58°59'W	Dec 85	4		geology
Harmony Point	62°19'S 59°14'W	Dec 85	3		geology

CHILE

BASE NAME	LOCATION	DATE OF INSTAL.	POP. sum.	win.	SCIENCE
ALL YEAR ROUND					
Capitan Arturo Prat (formerly Soberania)	62°30'S 59°41'W	6.2.47	45*	8	meteorology, oceanography
Bernado O'Higgins	63°19'S 57°54'W	8.2.48	~80*	29*	meteorology, seismology, human physiology
"Fildes Station"	62°11'S 58°55'W	29.5.85	~40		upper atmospheric physics, seismology, palaeobotanic fish studies, geological prospecting and oil exploration, pollution studies, geomorphology
Teniente Rodolpho Marsh/"Frei"	62°13'S 58°56'W	17.3.80 (7.3.69)	49**	57**	immunology, Air Force Base infrastructure, meteorology
SUMMER ONLY					
Gonzalez Videla	64°49'S 62°52'W	12.3.51	8		
Luis Carvajal***	67°46'S 69°55'W	Dec 83	10		meteorology
Yelcho	64°52'S 68°35'W	1962	~12		marine biology (occasional) (environmental data collection platform)
REFUGES					
Comodoro Guesalaga	67°46'S 68°45'W	28.2.62			
Cooper Mine	62°23'S 59°40'W	1956			
"Spring"	64°18'S 61°03'W	12.1.73			
"Ardley"	62°13'S 58°54'W	1986			
"Yankee Bay"	62°32'S 59°45'W	1952			
"Gutierrez"	62°57'S 60°35'W	1963			
"Los Gemelos"	63°22'S 57°34'W				
"Risopatron"	63°19'S 57°54'W	2.3.57			

* Mainly military.
** Includes military and families. Note: The Chilean complex at Marsh consists of three parts – Marsh (air facility), Frei (meteorology), and Fildes (scientific station).
*** Formerly the British base Adelaide, closed in 1977 and transferred to Chilean control in 1982.

Notes:
1. Chile has plans for two new winter bases south of the Antarctic circle, which will provide room for 100 summer personnel and 35 winterers.
2. In 1986/87, Chile announced plans to set up temporary camps on a trial basis in the vicinity of Deception Island (Padro Aguirre Cerda – 62°56'S 60°36'W), Elephant Island, Bridgeman Island, Paulet Island, Half Moon Island, Admiralty Bay, and Seymour Island.

CHINA

BASE NAME	LOCATION	DATE OF INSTAL.	POP. sum.	win.	SCIENCE
Chang Chen	62°13'S 58°58'W	20.2.85	~50	15	meteorology, geology, geophysics, geomorphology, surveying

* In 1984, China was "offered" a base site in the Ross Dependency by New Zealand, an offer that was politely declined in favour of the logistically easier – and certainly more crowded – King George Island.

FRANCE

BASE NAME	LOCATION	DATE OF INSTAL.	POP. sum.	win.	SCIENCE
ALL YEAR ROUND					
Dumont d'Urville	66°40'S 140°01'E	1.4.56	62*	28*	geomagnetics, seismology, meteorology, oceanography, upper atmosphere physics, cosmic ray physics, biology, pollution study, human physiology

* Summer maximum – 100. Winter maximum – 45.
Notes:
1. Sub-Antarctic bases maintained all year round on the islands of Kerguelen (49°21'S 70°12'E, first opened in January 1950), Crozet (46°26'S 51°52'E, 1961) and Amsterdam (37°50'S 77°34'E).
2. There are also unoccupied Antarctic refuges at Port Martin (66°49'S 141°23'E), and Charcot (69°22'S 139°02'E).

EAST GERMANY

BASE NAME	LOCATION	DATE OF INSTAL.	POP. sum.	win.	SCIENCE
ALL YEAR ROUND					
Schirmacher Oasis	70°46'S 110°14'E	Jan 76	9	6	meteorology, atmospheric chemistry, geophysics, geology, glaciology

WEST GERMANY

BASE NAME	LOCATION	DATE OF INSTAL.	POP. sum.	win.	SCIENCE
ALL YEAR ROUND					
Georg von Neumeyer	70°37'S 08°22'W	24.2.81	20	9	geophysics, geodesy, glaciology, meteorology, atmospheric physics, air chemistry and pollution, cosmochemistry
SUMMER ONLY					
Filchner	77°09'S 50°38'W	2.1.82	26 (max)		geophysics, glaciology, meteorology
Drescher	72°53'S 19°10'W	18.10.86	~15		geophysics, meteorology, glaciology
REFUGES					
Lillie Marlene	71°12'S 164°31'E	14.1.80	8		Joint American/German/NZ geophysical and geological research programmes including petrology and geochemistry
Gondwana	74°38'S	22.1.83	8		Joint American/German/NZ geophysical and geological research programmes including petrology and geochemistry

INDIA

BASE NAME	LOCATION	DATE OF INSTAL.	POP. sum.	win.	SCIENCE
ALL YEAR ROUND					
Dakshin Gangotri	70°05'S 12°00'E	Jan 84	~40	15	geology, geophysics, meteorology, upper atmospheric physics, cartography, marine biology

ITALY

BASE NAME	LOCATION	DATE OF INSTAL.	POP. sum.	win.	SCIENCE
SUMMER ONLY					
Terra Nova Bay	74°42'S 164°06'E	Dec 86	39		cosmic dust physics, meteorology, atmospheric physics, geology, geomorphology, geomagaphy, oceanography, biology, pollution studies

JAPAN

BASE NAME	LOCATION	DATE OF INSTAL.	POP. sum.	win.	SCIENCE
ALL YEAR ROUND					
Syowa	69°00'S 30°35'E	29.1.57	~40		meteorology, atmospheric physics, geophysics, geomagnetics, geochemistry
				29	seismology, geochemistry
Mizuho	70°42'S 44°20'E	21.7.70	~20		meteorology, glaciology*, geomagnetics
Asuka**	71°32'S 44°20'E	1.1.85	~12	8	geology, geophysics, meteorite search

* Glaciological traverse, Mizuho to Asuka, in 1986/87.
** Japan established a summer-only advance camp (74°21'S 34°59'E) in the summer of 1986/87, from which meteorological observations were conducted.

NEW ZEALAND

BASE NAME	LOCATION	DATE OF INSTAL.	POP. sum.	win.	SCIENCE
ALL YEAR ROUND					
Scott	77°51'S 166°46'E	Jan 57	~20*	12	meteorology, oceanography, ice physics, geomagnetics, upper atmosphere physics, seismology, marine biology, glaciology, geology, human physiology
SUMMER ONLY					
Vanda	77°31'S 161°28'E	Jan 67	3*		meteorology, hydrology, geochemistry, freshwater biology, geology
Cape Byrd	77°14'S 166°28'E	Jan 66	*		meteorology, ornithology, animal virology

* Precise summer population estimates are difficult. In 1986/87 there was a total of 83 field staff, of which about 30 were biologists operating out of Vanda and Cape Byrd, and 30 geologists operating out of north Victoria Land/Vanda station. The figures quoted in the table refer to base personnel only.
Note: New Zealand also maintains a winter station on Campbell Island (53°33'S 169°09'E) in the sub-Antarctic.

POLAND

BASE NAME	LOCATION	DATE OF INSTAL.	POP. sum.	win.	SCIENCE
ALL YEAR ROUND					
Henryk Arctowski	62°09'S 58°29'W	26.2.77	~25	19	meteorology, marine biology, geophysics, geology, human biology

SOUTH AFRICA

BASE NAME	LOCATION	DATE OF INSTAL.	POP. sum.	win.	SCIENCE
ALL YEAR ROUND					
SANAE (IV)	70°19'S 2°22'W	8.1.60	~70	15	auroral studies, airglow, cartography, cosmic rays, geodesy, geomagnetic geology, upper atmosphere meteorology, seismology
SUMMER ONLY					
Grunhoegna	72°02'S 2°48'W	Jan 83	16		geology, cartography, geodesy, geophysics
Sarie Marais	(820 km NW of SANAE)	Jan 87	12		geology, surveying

SOUTH AFRICA cont.

SUB-ANTARCTIC

BASE NAME	LOCATION	DATE OF INSTAL.	POP. sum.	win.	SCIENCE
Marion Island	46°52′S 37°52′E	29.12.47	12		entomology, geology, geomagnetics, limnology, biology, microbiology, volcanism
Gough Island	40°21′S 9°53′W	Jan 48	8		meteorology, biology

UNITED KINGDOM

BASE NAME	LOCATION	DATE OF INSTAL.	POP. sum.	win.	SCIENCE
ALL YEAR ROUND					
Halley	75°35′S 26°46′W	Jan 56	~30	19	meteorology, upper atmospheric physics, space plasma physics, ionospheric studies, atmospheric dynamics, radiation and chemistry (inc. ozone studies), geomagnetics, human physiology
Faraday	65°15′S 64°16′W	7.1.47	~24	12	meteorology, geomagnetics, seismology, space plasma physics, atmospheric dynamics, radiation and chemistry, human physiology, invertebrate biology
Signy	60°43′S 45°36′W	18.3.47	~28	12	inshore marine biology (fish), ornithology, microbiology, invertebrate biology, botany, freshwater biology and chemistry, environmental studies, human physiology
Rothera	67°34′S 68°07′W	late 76	~70*	12	geology, glaciology, geophysics, aeromagnetic surveys, cartography, glacio-chemistry, sedimentology and biostratigraphy, mineralization, meteorology, human physiology
SUMMER ONLY					
Damoy Point	64°49′S 63°31′W	14.11.75	*		air operations staging post into Rothera
Fossil Bluff	71°20′S 68°17′W	20.2.61	2		air operations refuelling post, meteorology
SUB-ANTARCTIC – ALL YEAR ROUND					
Bird Island	54°00′S 36°30′W	Nov 57	~9	3	ornithology, seal biology
SUMMER ONLY					
South Georgia	various since 1982		~6		biology field programmes at several different sites around the island

* Predominately field staff in transit.
Note: In recent years, the British Antarctic Survey have on occasion chartered a private ice-strengthened yacht to assist in geological and biological studies at South Georgia and on the Antarctic Peninsula.

URUGUAY

BASE NAME	LOCATION	DATE OF INSTAL.	POP. sum.	win.	SCIENCE
ALL YEAR ROUND					
Artigas	62°11′S 58°51′W	17.1.85	~20	10	meteorology, geomorphology, biology

UNITED STATES

BASE NAME	LOCATION	DATE OF INSTAL.	POP. sum.	win.	SCIENCE
ALL YEAR ROUND					
Amundsen-Scott	90°S	Jan 57	80+	17	meteorology, upper atmospheric physics, cosmic rays, glaciology, geomagnetics geodesy, solar radiation studies, atmospheric pollution monitoring, radio astronomy
McMurdo	77°51′S 166°40′E	Jan 56	950+	132	terrestrial and marine biology, geology, meteorology, upper atmospheric physics, glaciology, sea-ice studies, cosmic radiation, geophysics, petrology, geochemistry, geomorphology
Palmer	64°46′S 64°03′W	Jan 65 (completed 1968)	50+	7	meteorology, biology, VLF physics
Siple	75°56′S 84°15′W	Jan 69*	50+	7	meteorology, upper atmospheric physics, glaciology
SUMMER ONLY					
Byrd	80°01′S 119°32′W	Jan 57**	~8		meteorology, (mainly a fuel stop and weather station for aircraft operations)
"Dome C"	74°30′S 123°10′E	Nov 74	~20		meteorology, glaciology, (massive ice-drilling programme)
Beardmore*** South Camp	84°03′S 164°16′E	Jan 85	~100		geology, meteorology, glaciology, meteorities
Siple Coast	Eastern Ross Sea	1983	~55		glaciology, meteorology

* Siple rebuilt 1979/80. Closed 20 January 1984. Reopened November 1985. Closed January 1987. Due to reopen November 1987.
** Operated all year round until 1972. Rebuilt several times.
*** The site of a major international conference on the Antarctic Treaty System in January 1985.
Notes:
1. Summer programmes are also occasionally conducted at "Upstream B" (83°27′S, 139°39′W), Crary Ice Rise (83°45′S 166°05′W), Marble Point camp (77°25′S 163°40′E), Ellsworth Mountains camp (79°05′S 85°58′W), and camps in northern Victoria Land and in the Dry Valleys (especially Wright Valley). Each such camp is normally constructed to take up to 50 personnel.
2. In 1986/87 summer, four American biologists worked at Arctowski, the Polish station on King George Island, and several geologists supported by the Polar Duke worked on Seymour Island on the east coast of the Antarctic Peninsula.

USSR

BASE NAME	LOCATION	DATE OF INSTAL.	POP. sum.	win.	SCIENCE
ALL YEAR ROUND					
Molodezhnaya	67°40′S 45°51′E	14.1.63	390	129	meteorology, geomagnetics, geophysics, upper atmospheric physics (using rockets), aurorae, glaciology, geology, medicine
Mirny	66°33′S 93°01′E	13.2.56	90	67	meteorology, geophysics, seismology, glaciology medicine, climatology
Vostok	78°28′S 106°48′E	16.12.57	37	29	glaciology, meteorology, geosciences, auroral physics
Novolazarevskaya	70°46′S 11°50′E	18.2.61	60	57	geosciences, meteorology, glaciology
Bellingshausen	62°12′S 58°56′W	22.2.68	35	25	meteorology, biology, geosciences, environmental impact
Leningradskaya	69°30′S 159°23′E	25.2.71	18	18	meteorology, geosciences
Russkaya	74°46′S 136°51′W	9.3.80	12	12	meteorology, medicine
SUMMER ONLY					
Druzhnaya III	Cape Norvegica (nr SANAE)	Dec 86	12		geology, meteorology
Druzhnaya II	74°30′S 62°00′W	Nov 80	29		geology, meteorology
Soyuz	70°35′S 68°47′E	Nov 82 (opened 86/87)	17		meteorology, geology
OCCASIONAL					
Oaziz	66°16′S 100°45′E	15.10.56 (reopened 86/87)	5		meteorology, geology
Komsomolskaya (winter party 1987)	74°06′S 97°28′E	6.2.57 (reopened 86/87)	37	12	glaciology, geosciences

Notes:
1. Komsomolskaya has been reopened after 28 years of inactivity, and is provisionally renamed "Progress".
2. Druzhnaya I was lost, along with the Argentine Base Belgrano I, in July 1986, when a large amount of ice calved off the Filchner Ice Shelf (see Argentina). Druzhnaya I was due to be reopened in October 1986 in support of geology programmes in the Pensacola Mountains.
3. Russkaya was operated as a makeshift summer camp from 1973, but a permanent installation was not successfully deployed until November 1982.
4. Dobrowolski has been reclaimed by the USSR from Poland, without negotiation, and renamed Oaziz.
5. In 1986/87, the USSR also operated an extra summer station in the Larsemann Hills near Prydz Bay (about 76°E); population approx 10. Druzhnaya III was established by the Kapitan Kondrat'yev, Soyuz (edge of Amery Ice Shelf), and the summer station near Prydz Bay by the Vasiliy Fedoseyev.

ANTARCTIC BASES

This map shows the positions of the 68 bases present in Antarctica together with the countries responsible for them. Of the total number, 42 are manned all year round and 26 are occupied only during the austral summer months.

KEY

ARG	= Argentina	IN	= India
AUS	= Australia	NZ	= New Zealand
BR	= Brazil	PL	= Poland
CHI	= China	SA	= South Africa
CHL	= Chile	UK	= United Kingdom
DDR	= East Germany	UR	= Uruguay
FR	= France	USA	= United States
FRG	= West Germany		of America
IT	= Italy	USSR	= Soviet Union
JAP	= Japan		

* summer only bases

COMMANDANTE FERRAZ (BR)
ARCTOWSKI (PL)
TENIENTE JUBANY (ARG)
ARTIGAS (UR)
ASTRONOMO CRULS (BR)*
TENIENTE RODOLFO MARSH (CHL)
BELLINGSHAUSEN (USSR)
CAPITAN ARTURO PRAT (CHL)
GREAT WALL (CHI)

GENERAL BERNARDO O'HIGGINS (CHL)
ESPERANZA (ARG)
VICE COMMODORO MARAMBIO (ARG)
GONZALES VIDELA (CHL)
PRIMAVERA (ARG)*
MATEINZO (ARG)*
YELCHO (CHL)
DAMOY POINT (UK)*
FARADAY (UK)
PALMER (USA)
GENERAL SAN MARTYN (CHL)
ROTHERA (UK)
TENIENTE CARVAJAL (CHL)*

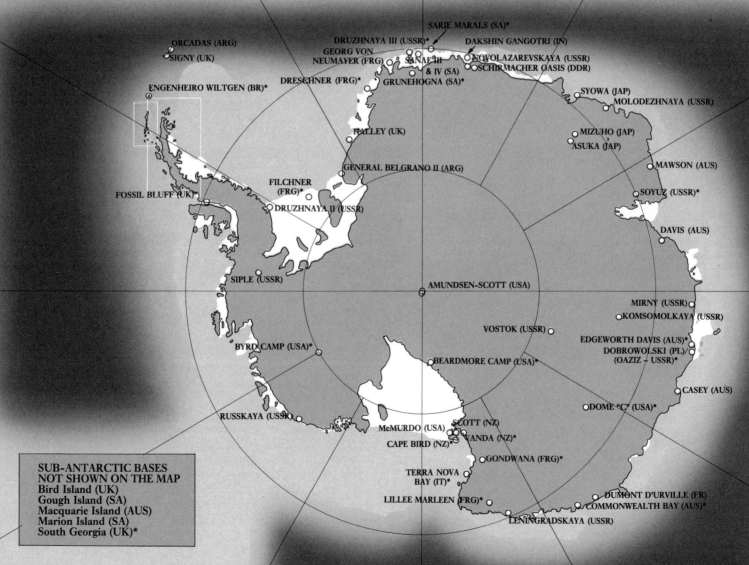

SARIE MARALS (SA)*
DRUZHNAYA III (USSR)*
DAKSHIN GANGOTRI (IN)
GEORG VON NEUMAYER (FRG)
SANAE III & IV (SA)
NOVOLAZAREVSKAYA (USSR)
SCHIRMACHER OASIS (DDR)
ORCADAS (ARG)
SIGNY (UK)
ENGENHEIRO WILTGEN (BR)*
DRESCHNER (FRG)*
GRUNEHOGNA (SA)*
SYOWA (JAP)
MOLODEZHNAYA (USSR)
HALLEY (UK)
MIZUHO (JAP)
ASUKA (JAP)
MAWSON (AUS)
GENERAL BELGRANO II (ARG)
SOYUZ (USSR)*
FILCHNER (FRG)*
FOSSIL BLUFF (UK)
DRUZHNAYA II (USSR)
DAVIS (AUS)
SIPLE (USSR)
AMUNDSEN-SCOTT (USA)
MIRNY (USSR)
KOMSOMOLKAYA (USSR)
VOSTOK (USSR)
EDGEWORTH DAVIS (AUS)*
DOBROWOLSKI (PL)/ (OAZIZ – USSR)*
BYRD CAMP (USA)*
BEARDMORE CAMP (USA)*
CASEY (AUS)
DOME "C" (USA)*
RUSSKAYA (USSR)
McMURDO (USA)
SCOTT (NZ)
CAPE BIRD (NZ)*
VANDA (NZ)*
GONDWANA (FRG)*
DUMONT D'URVILLE (FR)
TERRA NOVA BAY (IT)*
COMMONWEALTH BAY (AUS)*
LILLEE MARLEEN (FRG)*
LENINGRADSKAYA (USSR)

SUB-ANTARCTIC BASES NOT SHOWN ON THE MAP

Bird Island (UK)
Gough Island (SA)
Macquarie Island (AUS)
Marion Island (SA)
South Georgia (UK)*

ANTARCTIC TREATY:
Membership & Summary

Several nations have recently demonstrated growing interest in Antarctica. Most of them have already acceded to the Treaty, and may introduce substantial Antarctic science programmes in the near future, with a view to attaining full Consultative Status. It appears that this process of ratification, installation of stations and full Consultative Status, is accelerating within the international community, despite criticism from developing countries that the Antarctic Treaty System is not properly representative of the lofty principle of the Common Heritage of Mankind. Although the Antarctic Treaty System is in favour of enlarging the membership, an excessively large number of countries presents a worrying bureaucratic problem: the chances of achieving a workable consensus on important issues is greatly diminished.

The Netherlands acceded to the Treaty on 30 March 1967, and shared in four Antarctic expeditions with Belgium in the 1960s. Low level interest has been maintained since through the Netherlands Academy of Sciences. Formal interest was renewed in December 1984, when the Ministry of Education and Science decided to allocate 1 million Dutch guilders to support young scientists wishing to participate in Antarctic research. There are plans to obtain a ship for oceanographic research in the South Atlantic and Southern Oceans. In 1986/87, Holland contributed to a geoseismic cruise in *Polarstern*.

Peru acceded to the Treaty on 10 April 1981. In 1984, a Peruvian scientist wintered at the Argentine base of San Martin. In 1982/83, Peru sent observers south with Australia, Chile and Brazil. Peru is currently preparing an Antarctic Research Programme. Peru is attracted by the Brazilian "frontage theory" for Antarctic sovereignty, which would give Peru territory in Antarctica at the expense of bitter rival, Chile.

Spain acceded to the Treaty on 31 March 1982 (coincidentally, two days before the Falkland Islands war) and despatched its first Antarctic expedition rather haphazardly in the 1982/83 austral summer season. The research vessel *Ideus de Mazo* (a tourist yacht) cruised off the Antarctic Peninsula in February 1983. It was a private expedition designed to influence public opinion and stimulate government support, which it did. Representatives of the Spanish Navy, the Marine Ministry and the Institute of Oceanography were among 23 expedition members. The expedition is now considered as the first step towards a permanent Spanish base. In 1983 Spain announced to the UN Study on Antarctica her desire to achieve full Consultative Status.

Sweden acceded to the Treaty on 24 April 1984, after a long absence from involvement in the region. Sweden was one of the first countries to send an expedition to Antarctica during the "Heroic Age" (Nordenskjold, 1901-4), and participated in the British-Norwegian-Swedish expedition of 1949-52. Sweden and New Zealand have issued a joint communique that the two countries will cooperate actively in polar science, possibly under Treaty. Two Swedish scientists worked at Scott base in 1984/85. Sweden has established a Polar Research Secretariat and expects to send more scientists south in 1987/88, with the German programmes.

Cuba became the 32nd Treaty nation on 16 August 1984. Two Cuban geographers, participating in the Soviet Antarctic Expedition of 1982/83, raised the Cuban flag alongside the Soviet and East German flags at the Soviet station, Molodezhnaya. Their initiative had the express approval of Fidel Castro, whose attention was drawn to the region's political importance by the occurrence of the Falkland Islands war. Several Cuban scientists have since gained valuable experience with the Soviets, fuelling speculation that Cuba might mount her own independent expedition to Antarctica, and substantiate a claim for full Consultative Status. Her vote within the Antarctic Treaty System would be used in support of the Soviet Union.

South Korea acceded to the Treaty on 28 November 1986, after an advance party of nine scientists carried out a rudimentary science programme in the summer of 1985/86, with the intention of selecting a suitable site for a station on King George Island: the urbanization of King George Island is likely to continue unabated. In the same season, a party of seven climbers mounted an expedition to the Ellsworth Mountains. Both parties were supported logistically by the Chilean Air Facility on King George Island. In April 1987, eight Korean scientists returned to King George Island, again with Chilean assistance, and chose a site for a future scientific station.

South Korea has substantial fishing interests around Antarctica. Her first krill catching expedition worked off the coast of Enderby Land in 1983/84.

North Korea acceded to the Treaty on 21 January 1987, barely two months after South Korea.

Austria acceded to the Antarctic Treaty in August 1987.

Ecuador acceded in September 1987.

Probable developments

It is possible that the Antarctic Treaty Consultative Party Membership could increase from its current 20 to 27 by 1990 (possible new Consultative Parties are: North and South Korea, Cuba, Sweden, the Netherlands, Spain and Peru).

It is also expected that the following countries will accede to the Antarctic Treaty in due course: Switzerland (for glaciological research) and Canada (as the only country of the top seven economic nations not to do so and with considerable, indeed unrivalled, polar expertise).

So far, there is no sign of representation from the Middle East, although Saudi Arabia has devoted considerable time and effort (not to mention money), on research into iceberg utilization for fresh water, and may therefore consider acceding at some point. Whether this would produce a "domino" effect, encouraging the more moderate Arab states to accede to the Treaty, is uncertain. No black African nations have acceded, mainly because of the presence of South Africa as an original signatory. Without South Africa, Mozambique and Angola would almost certainly accede.

Therefore, it is also possible that the total number of nations acceding to the Antarctic Treaty System may increase to about 40 by the year 1990.

Commendation and condemnation
Other nations so far not considered (for instance, Mexico, Bolivia, Singapore, Paraguay and Thailand) have in general commended the principles of the Antarctic Treaty System, whilst asking for greater openness and more international involvement. Such nations may well feel motivated to accede in due course and (like Greece, in January 1987) without much warning.

Most small members of the Non-Aligned Movement, including much of Africa, and some countries in Southern Asia and the Middle East are quite vitriolic in their condemnation of the Antarctic Treaty System, especially as the richer Antarctic nations are close to completing a Treaty for the exploitation of Antarctic mineral resources. Such nations are strong advocates of the Common Heritage of Mankind principle for Antarctica, within the UN.

Changes of standpoint
The situation within each country can be quite mercurial, however. For instance, Nigeria has followed Antarctic developments with particular interest, and supports the Common Heritage of Mankind approach. But if the tendency for membership within the Antarctic Treaty System was suddenly to turn from a trickle into a rush, Nigeria might be prepared to change policy in order to avoid missing out on any mineral benefit from the region.

Like Saudi Arabia in the Middle East, Nigeria could also trigger a domino effect among the countries in Africa, in much the same way, for instance, that the entrance of Brazil into Antarctic politics has triggered sympathetic responses in Uruguay, Peru and Ecuador.

ANTARCTIC TREATY MEMBERSHIP

+ *United Kingdom*	*31.5.60*	Bulgaria	11.9.78
+ *South Africa*	*21.6.60*	+ Germany, BRD	5.2.79 (3.3.81)
+ *Belgium*	*26.7.60*	+ Uruguay	11.1.80 (7.10.85)
+ *Japan*	*4.8.60*	Papua New Guinea*	16.3.81
+ *United States of America*	*18.8.60*	+ Italy	18.3.81 (5.10.87)
+ *Norway*	*24.8.60*	Peru	10.4.81
+ *France*	*16.9.60*	Spain	31.3.82
+ *New Zealand*	*1.11.60*	+ China, Peoples' Republic	8.6.83 (7.10.85)
+ *Soviet Union*	*2.11.60*	+ India	19.8.83 (12.9.83)
+ Poland	8.6.61 (29.7.77)	Hungary	27.1.84
+ Argentina	23.6.61	Sweden	24.4.84
+ Australia	23.6.61	Finland	15.5.84
+ Chile	23.6.61	Cuba	16.8.84
Czechoslovakia	14.6.62	Korea (Seoul)	28.11.86
Denmark	20.5.65	Greece	8.1.87
Netherlands	30.3.67	Korea (Pyongyang)	21.1.87
Romania	15.9.71	Austria	25.8.87
+ Germany, DDR	19.11.74 (5.10.87)	Ecuador	15.9.87
+ Brazil	16.5.75 (12.9.83)		

Original signatories; the 12 states which initialled the Treaty on 1 December 1959, are *italicized*; the dates given are those of the deposition of the ratifications of the Treaty.

+ Consultative Parties; 20 states, the 12 original signatories and 8 others which achieved this status after becoming actively involved in Antarctic Research (with dates in brackets).

* Papua New Guinea succeeded to the Treaty after becoming independent of Australia.

ANTARCTIC TREATY: SUMMARY OF BASIC PROVISIONS

ARTICLE I Antarctica shall be used for peaceful purposes only. All military measures, including weapons testing, are prohibited. Military personnel and equipment may be used, however, for scientific purposes.

ARTICLE II Freedom of scientific investigation and cooperation shall continue.

ARTICLE III Scientific program plans, personnel, observations and results shall be freely exchanged.

ARTICLE IV The treaty does not recognize, dispute, or establish territorial claims. No new claims shall be asserted while the treaty is in force.

ARTICLE V Nuclear explosions and disposal of radioactive wastes are prohibited.

ARTICLE VI All land and ice shelves below 60° South Latitude are included, but high seas are covered under international law.

ARTICLE VII Treaty-state observers have free access—including aerial observation—to any area and may inspect all stations, installations, and equipment. Advance notice of all activities and of the introduction of military personnel must be given.

ARTICLE VIII Observers under Article VII and scientific personnel under Article III are under the jurisdiction of their own states.

ARTICLE IX Treaty states shall meet periodically to exchange information and take measures to further treaty objectives, including the preservation and conservation of living resources. These consultative meetings shall be open to contracting parties that conduct substantial scientific research in the area.

ARTICLE X Treaty states will discourage activities by any country in Antarctica that are contrary to the treaty.

ARTICLE XI Disputes are to be settled peacefully by the parties concerned or, ultimately, by the International Court of Justice.

ARTICLE XII After the expiration of 30 years from the date the treaty enters into force, any member state may request a conference to review the operation of the treaty.

ARTICLE XIII The treaty is subject to ratification by signatory states and is open for accession by any state that is a member of the UN or is invited by all the member states.

ARTICLE XIV The United States is the repository of the treaty and is responsible for providing certified copies to signatories and acceding states.

BIBLIOGRAPHY

PRINCIPAL SOURCES

Antarctica: Great Stories from the Frozen Continent, Reader's Digest, NSW/Australia, 1985

Bonner, W. N. and Walton, D. W. H. (eds). *Antarctica*, Pergamon Press, Oxford, 1985

Brewster, Barney. *Antarctica: Wilderness at Risk*, A. H. & A. W. Reed Ltd, Wellington, New Zealand, 1982

King, H. G. R. *The Antarctic*, Blandford Press, London, 1969.

Pyne, Stephen J. *The Ice: A Journey to Antarctica*, University of Iowa Press, Iowa City, 1986

Schultess, Emil. *Antarctica*, Collins, London, 1961

Shapley, Deborah. *The Seventh Continent: Antarctica in a Resource Age*, Resources for the Future Inc, Washington DC, 1985

Walton, D. W. H. (ed). *Antarctic Science*, Cambridge University Press, Cambridge, 1987

GENERAL SOURCES

Allen, Oliver E. *Atmosphere*, Time-Life Books, Amsterdam, 1983

Antarctica: A Continent for Science, British Antarctic Survey and the Natural Environment Research Council,

Bailey, Ronald H. *Glacier*, Time-Life Books, Amsterdam, 1982

Barnes, James N. *Let's Save Antarctica*, Greenhouse Publications, Victoria, Australia, 1982

Beck, Peter J. *The International Politics of Antarctica*, Croom Helm, 1986

Benninghoff, W. S. and Bonner, W. N. *Man's Impact on the Antarctic Environment*, Scientific Committee on Antarctic Research, Scott Polar Research Institute, Cambridge, 1985

Biologist: *Special Issue: Antarctica*, Vol. 32, No. 3, June 1985, Institute of Biology, London

Bonner, W. N. and Lewis Smith, R. I. *Conservation Areas in the Antarctic*, SCAR, March 1985, Scott Polar Research Institute, Cambridge

Byrd, Adm. Richard E. *Alone*, Queen Anne Press/Macdonald & Co. London, 1938/new Edn 1987

Chester, Jonathan. *Going to Extremes*, Doubleday Australia, NSW/Australia, 1986

Couper, Alistair (ed). *The Times Atlas of the Oceans*, Times Books, London, 1983

Dooling, Dave. Satellite Data Alters View on Earth-Space Environment, *Spaceflight*, Vol. 29, July 1987

Drewry, D. J. *Antarctica: Glaciological and Geophysical Folio*, Scott Polar Research Institute, Cambridge, 1983

Drewry, David J. Antarctica Unveiled, *New Scientist*, 22 July, 1982

Eastman, Joseph T. and DeVries, Arthur L. *Antarctic Fishes*, Scientific American, November 1986

Foster, James. Optics of the snow and sky, *Antarctic Journal*, December 1985, National Science Foundation, Washington DC

Frank, L. A., Craven, J. D. and Rairden R. L. *Images of the Earth's Aurora and Geocorona from the Dynamics Explorer Mission*, Adv. Space. Res, Vol. 5, No. 4, pp 53–68, 1985

Frank, L. A., Craven, J. D. et al. The Theta Aurora, *Journal of Geophysical Research*, Vol. 91, No. A3, March 1, 1986

Gow, Anthony J. *Preliminary Results of Studies of Ice Cores from the 2164m Deep Drill Hole, Byrd Station, Antarctica*, Reprint from ISAGE Symposium, Hanover, USA, 3–7 September 1968

Gow, Anthony J. Relaxation of Ice in Deep Drill Cores from Antarctica, *Journal of Geophysical Research*, Vol. 76, No. 11, 10 April 1971

Gow, Anthony J. and Williamson, Terence. *Volcanic Ash in the Antarctic Ice Sheet and its Possible Climatic Implications*, Earth and Planetary Science Letters 13 (1971) pp 210–218, North-Holland Publishing Co

Greene, S. W. et al. *Terrestrial Life of Antarctica*, Folio 5, Antarctic Map Series, American Geophysical Society, New York, 1967

Greenpeace/ASOC. *ECO Newsletters and Background Papers*, 1983–87

Keys, Harry. *Towards a New Shape Classification of Antarctic Icebergs*, Iceberg Research, No. 12, April 1986

Laws, Richard M. Antarctica: A convergence of life, *New Scientist*, 1 September 1983

Laws, Richard M. The Ecology of the Southern Ocean, *American Scientist*, Vol. 73, Jan–Feb, 1985

Lucchitta, Baerbel K. et al. *Multispectral Landsat Images of Antarctica*, US Geological Survey Bulletin, 1696, Washington DC 1986

McWhinnie, Mary A. and Denys, Charlene J. The High Importance of the Lowly Krill, *Natural History*, March 1980, pp 66–73

Maran, Stephen P. Rocks From Mars, *Natural History*, November 1983

Marvin, Ursula B. Extraterrestrials have Landed on Antarctica, *New Scientist*, 17 March 1983

Perry, Richard. *The Polar Worlds*, David & Charles, Newton Abbot, 1973

Polar Regions Atlas. Central Intelligence Agency, May 1978

Project Antarctica. *Expedition Notebook*, Society Expeditions, Seattle, Washington

Rairden, R. L., Frank, L. A. and Craven, J. D. Geocoronal Imaging with Dynamics Explorer, *Journal of Geophysical Research*, Vol. 91, No. A12, December 1986

Reuning, Winifred M. *Antarctic Journal of the United States*, 1985 Review, Vol XIX – No 5, National Science Foundation, Washington DC

Rycroft, Michael. A View of the Upper Atmosphere from Antarctica, *New Scientist*, 28 November 1985

Shackleton, Sir Ernest. *South*, Century Hutchinson, London, 1919/new Edn 1983

Shackleton, Keith. *Ship In the Wilderness*, J. M. Dent, London, 1986

Schoeberl, Mark R. and Krueger, Arlin J. *The Morphology of Total Ozone as seen by TOMS*, Geophysical Res. Lett., 13, No. 12, pp 1217–1220, November 1986

Sparks, John and Soper, Tony. *Penguins*. David & Charles, Newton Abbot, 1967, 1987

Stolarski, R. S., Krueger, A. J. et al.

Nimbus 7 Satellite Measurements of the Springtime Antarctic Ozone Decrease, Nature, 332, N. 6082, pp 808–811

Stonehouse, Bernard. *Sea Mammals of the World,* Penguin Books, 1985

Swithinbank, Charles. A Year with the Russians in Antarctica, *The Geographical Journal,* Vol. 132, Part 4, December 1966

Swithinbank, Charles. To the Valley Glaciers that Feed the Ross Ice Shelf, *The Geographical Journal,* Vol. 130, Part 1, March 1964

Thomson, Michael and Swithinbank, Charles. The Prospect for Antarctic Minerals, *New Scientist,* 1 August 1985

Watson, Lyall. *Sea Guide to Whales of the World,* Hutchinson, London, 1981

Weyant, W. S. *The Antarctic Atmosphere: Climatology of the Surface Environment: Folio 8,* Antarctic Map Series, American Geographical Society, 1967

Wolff, Eric. The Answer Lies In The Ice, *The Geographical Magazine,* February 1987

Wolfrum, Rüdiger (ed). *Antarctic Challenge II: Proceedings of an Interdisciplinary Symposium,* September 17th–21st, 1985, Duncker & Humblot, Berlin, 1986

Zwally, H. J. et al. *Antarctic Sea Ice, 1973–1976.* NASA, Washington DC, 1983

Zwally, H. J. *Observing Polar-Ice Variability.* Annals of Glaciology 5, International Glaciological Society, 1984

Zwally, H. J. et al. Variability of Antarctic Sea Ice and Changes in Carbon Dioxide, *Science,* Vol. 220, No. 4601, 3 June 1983

OTHER SOURCES CONSULTED

Daily Telegraph
Discover
The Guardian
International Herald Tribune
The Independent
The Listener
New Scientist
Newsweek
The Observer
Science Digest
The Sunday Times
Time
The Times

ABBREVIATIONS

ANARE	Australian National Antarctic Research Expedition
ASOC	Antarctic and Southern Oceans Coalition
ATCM	Antarctic Treaty Consultative Meetings
ATCP	Antarctic Treaty Consultative Party
ATS	Antarctic Treaty System
BAS	British Antarctic Survey
BIOMASS	Biological Investigation of Marine Antarctic Systems and Stocks
CCAMLR	Convention for the Conservation of Antarctic Marine Living Resources
CCAS	Convention for the Conservation of Antarctic Seals
ESMR	Electrically Scanning Microwave Radiometer
FIBEX	First International Biological Experiment
GARP	Global Atmospheric Research Programme
IGY	International Geophysical Year
ISTP	International Solar Terrestrial Physics Programme
IUCN	International Union for the Conservation of Nature and Natural Resources
IWC	International Whaling Commission
MV	Motor Vessel
NASA	National Aeronautics and Space Administration
NGO	Non-Governmental Organisation
NOAA	National Oceanographic and Atmospheric Administration

RRS	Royal Research Ship
RV	Research Vessel
SAR	Synthetic Aperture Radar
SCAR	Scientific Committee for Antarctic Research
SIBEX	Second International Biological Experiment
SMA	Specially Managed Area
SPA	Specially Protected Area
SS	Steam Ship
SSSI	Site of Special Scientific Interest
TOMS	Total Ozone Mapping Spectrometer
UN	United Nations
cm	centimetres
cu	cubic
ft	feet
g	grammes
gal	gallons
in	inches
kg	kilogrammes
km	kilometres
l	litres
lb	pounds (weight)
m	metres
fg	microgrammes
mls	miles
ml	millilitres
mm	millimetres
mya	million years ago
oz	ounces
sec	second
sq	square
yd	yards

GAZETTEER

The following gazetteer contains all the names that appear on the map of Antarctica on pages 66 and 67. For the locations of the Antarctic bases, see the map on page 181.

Abbot Ice Shelf	B3	Hercules Dome	C3	South Orkney Islands	A1
Alexander Island	B2	Horlick Mountains	C3	South Shetland Islands	A1
Allan Hills	B2			Spaatz Island	B2
Amery Ice Shelf	E2	Kemp Land	E1		
Amundsen Sea	A3	King Haakon VII Sea	D1	Terre Adélie	E4
Antarctic Peninsula	B2			Thorshavnheiane	D1
Anvers Island	A1	Lake Vanda	D4	Thurston Island	B3
		Lambert Glacier	E2	Titan Dome	D3
Balleny Islands	D5	Larsen Ice Shelf	B2	Transantarctic Mountains	C3
Beardmore Glacier	D3	Law Dome	E3		
Bellingshausen Sea	A2			Usarp Mountains	D4
Berkner Island	C2	Mac. Robertson Land	E2		
Britannia Range	D3	Marie Byrd Land	C3	Valkyrjedomen	D2
Bunger Hills	F3	Maudheimvidda	C1	Vestfold Hills	E2
Byrd Glacier	D3	McMurdo Sound	D4	Victoria Land	D4
		Mount Erebus	D4	Vinson Massif	C2
Cape Adare	D4	Mount Tyree	B2		
Cape Hallett	D4	Mulock Glacier	D4	Weddell Sea	C1
Carney Island	B3			West Ice Shelf	F2
Coats Land	C1	Nimrod Glacier	D3	Wilhelm II Land	E3
Colvocoresses Bay	E3			Wilkes Land	E3
Commonwealth Bay	E4	Oates Land	D4		
Dome Argus	D2	Palmer Land	B2		
Dome Circe	E3	Pensacola Mountains	C2		
Dronning Fabiolafjella	E1	Port Martin	E4		
Dronning Maud Land	D2	Prince Charles Mountains	E2		
Dry Valleys	D4	Princess Elizabeth Land	E2		
Dumont d'Urville Sea	E4	Prydz Bay	E2		
Ekström Ice Shelf	C1	Queen Mary Land	E3		
Ellsworth Land	B2	Queen Maud Mountains	C3		
Ellsworth Mountains	C2				
Enderby Land	E1	Riiser-Larsenisen	C1		
Evans Ice Stream	B2	Ronne Ice Shelf	C2		
		Roosevelt Island	C3		
Ferrar Glacier	D4	Ross Ice Shelf	C3		
Filchner Ice Shelf	C2	Ross Island	D4		
Fimbulheimen	D1	Ross Sea	C4		
Fimbulisen	D1				
Framnes Mountains	E2	Scott Glacier	F3		
Foundation Ice Stream	C2	Sentinel Range	C2		
		Seymour Island	B1		
George V Land	D4	Shackleton Glacier	C3		
George VI Ice Shelf	B2	Shackleton Ice Shelf	F3		
Getz Ice Shelf	B3	Shackleton Range	C2		
Graham Land	B2	Siple Dome	C3		
Grove Mountains	E2	Siple Island	B3		

INDEX

Page numbers in **bold** refer to illustrations and captions

AUTHOR'S ACKNOWLEDGMENTS

A complex book such as this could not have been completed without the hard work and generous assistance of a wide range of people, all of whose contributions are gratefully acknowledged by the author.

Firstly, I am grateful to the contributors to this volume, the fine photographers whose credits appear below, especially Doug Allan, Colin Monteath and Eliot Porter whose work we are privileged to present. Many thanks also to Sir Peter Scott and Martin Baker (whose own book on Antarctica will be published by the time you read this note) and Jim Barnes whose organization, the Antarctic and Southern Ocean Coalition, can be contacted at 218 D. Street SE, Washington DC 20003, USA.

For efforts over and above the call of duty my heartfelt thanks to Ian Whitelaw, Jane Laing and Alex Arthur for making this book read coherently and look beautiful. It has become as much their book as mine over the long months of work, and they have all contributed immensely to its final appearance.

At Greenpeace, my greetings to all at International and other offices around the world. Special thanks to campaigners Roger Wilson and Kelly Rigg for their patience and encouragement, Maj de Poorter and the over-winterers for their writings, Jacqueline Geering and Jay Townsend for photographic assistance, Leslie Busby and Sabine Koch for text checking, Peter Bahouth and Nick Gallie for solidarity in the face of adversity.

Special thanks to Martin Leeburn (for his constant support and troubleshooting), David McTaggart (for giving me inspiration, friendship and opportunities) and Cornelia Durrant (without whom I wouldn't be here).

My Antarctic education was gained with the kind assistance of many people at the Scott Polar Research Institute and the British Antarctic Survey. In particular I wish to thank Bob Headland and Charles Swithinbank, who allowed me access to their vast knowledge and explained patiently the errors of my ways. Any mistakes that remain are my responsibility.

In addition, I would like to acknowledge the help of Chris Gilbert, Dr Drewry, Nigel Bonner, Dr W. Block, David Rootes, Dr D. W. S. Limbert, Mrs G. Joan Smith, Dr Peter D. Clarkson and Bernard Moran.

At Dorling Kindersley, I would like to thank Christopher Davis, Jackie Douglas and Roger Bristow who, together with our agent Jeffrey Simmons, made it practically possible to undertake this complex project.

Thanks are also due to: John Cleare of the Mountain Camera Picture Library for his useful introductions; Dr Jorn Sievers of the Institut für Angewandte Geodäsie; Stephen Knight; Edwin Mickleburgh; Winifred Reuning of the National Science Foundation; Cassandra Phillips; Mike Marten and Rosemary Taylor of the Science Photo Library; and Paul Mathews of Stanley Gibbons.

Lastly but most importantly, my love and thanks to Tanya Seton, Alexi and Louis who are always with me through thick and thin.

This book is dedicated to Greenpeacers worldwide in the hope that it will provide encouragement and inspiration in their struggle to achieve a peaceful and pollution-free planet.

PHOTOGRAPHIC CREDITS

The following organizations and individuals provided photographs and gave permission for them to be reproduced.

Alfred-Wegener-Institut für Polar und Meeresforschung: 57
Doug Allan: half-title page, 9 (inset top), 14/15, 30, 35, 36/37 (top), 37 (bottom), 40 (right), 41, 47 (bottom), 49 (top), 59 (bottom left), 65, 68 (inset), 68/69 (main picture), 76/77, 78/79, 82, 84, 85, 88 (top), 91, 92, 93, 95, 96/97, 98, 99, 100/101, 124, 140, 143, 158, 159, 161 (top)
Peter Breen: 133 (top)
British Antarctic Survey: 58 (top); *Dr W. Block:* 73; *W. N. Bonner:* 145 (top); *J. C. Ellis-Evans:* 70/71 (main picture); *Inigo Everson:* 81, 146, 147 (bottom); *S. Fraser:* 7 (inset), 113, 125 (bottom); *C. J. Gilbert:* front cover, endpapers, 12/13 (main picture), 54, 88 (bottom), 150/151 (bottom); *P. Gurling:* 125 (top); *L. D. B. Herrod:* 11 (inset left), 108/109 (main picture); *R. M. Laws:* 9 (inset bottom), 147 (top); *R. I. Lewis-Smith:* 94; *D. Macdonald:* 31; *A. Milne:* 59 (top); *A. B. Moyes:* 28; *R. G. Renner:* 58 (bottom); *Brian Thomas:* 149; *M. R. A. Thomson:* 17; *Trans-Antarctic Expedition:* 151 (top); *W. Vaughan:* 145 (centre); *A. Ward:* 59 (bottom right); *D. White:* 55, 86/87
Cambridge Paleomap Services Ltd – courtesy of Dr A. G. Smith: 16
Jonathan Chester: 34, 44, 53 (left), 86 (left), 122
Cold Regions Research and Engineering Laboratory, Snow and Ice Branch – courtesy of Anthony J. Gow: 27 (right)
Greg Crocker: 112 (right)
Environmental Research Institute of Michigan: 18
Greenpeace: 5, 11 (inset right), 13 (inset), 112 (left), 132/133 (bottom), 136, 137, 156 (inset), 157 (main picture), 160, 161 (bottom), 162, 163, 164, 165, 166, 167, 169, 170, 172, 173, 175, back cover
Institut für Angewandte Geodäsie: satellite images – Landsat-5 mss

data purchased from EOSAT; data digitally processed by Institut für Angewandte Geodasie, Frankfurt am Main: 33, 38, 39
Bob McKerrow: 133 (centre), 135, 142
Colin Monteath: title page (main picture), 8/9 (main picture), 10/11 (main picture), 45, 51 (bottom), 71 (top), 72, 105, 108 (inset), 115, 116, 121 (bottom), 123, 126, 128 (top), 134, 138, 139
NASA: title page (inset), 63 (top and bottom)
NASA/Goddard Space Flight Center – courtesy of H. Jay Zwally, Oceans and Ice Branch, Laboratory for Oceans: 42/43; *courtesy of Dr Arlin J. Krueger, Toms sensor scientist, Atmospheric Chemistry and Dynamics Branch:* 61
National Oceanic and Atmospheric Administration: 52
National Science Foundation: Ann Hawthorn: 53 (right), 118, 129; *Russ Kinne:* 63 (centre), 127 (bottom); *Cornelius Sullivan:* 40 (bottom left)
Ian Paterson: 121 (top)
Eliot Porter: 6/7 (main picture), 50, 51 (top)
Royal Geographical Society: 110/111
Charles Swithinbank: 20, 29, 47 (top), 56/57, 127 (top and centre), 128 (centre), 130, 131, 153, 155
University of Iowa, Department of Physics and Astronomy – courtesy of Professors L. A. Frank and J. D. Craven: 64
US Geological Survey, EROS Data Center, Sioux Falls: 27 (bottom left), 46, 128 (bottom)
US Geological Survey, Flagstaff, Arizona – courtesy of Baerbel K. Lucchitta: 24/25, 26, 32, 48/49, 62
US Navy: 37 (right)

ILLUSTRATORS *Martin Camm, Brian Delf, Eugene Fleury, Nicholas Hall and Janos Marffy*

PASTE-UP *Patrizio Semproni*